timber—the sugar tree, bass, oak, hickory, ash, &c. It is estimated that about one-fifth of these forests is occupied by pine; two-fifths with sugar, oak, bass, hickory, &c., and two-fifths with tamarack, cedar, alder, black ash, white maple, and aspen.

The lumbering business will doubtless take the lead, on and along these rivers for many years; but whenever the pine shall be exhausted, or lumber cease to bear remunerative prices, the more arable lands will be called into requisition, for agricultural purposes, and sustain a more dense population than any country of prairie and openings can do. In fact, as before remarked, the two pursuits are already thriving to some extent side by side in immediate proximity; many men finding it quite as profita- ble to raise grain, grasses and potatoes, with which to furnish those felling the pine forests, as to manufacture boards. Did time and space permit, this position could be verified by facts in detail, by citing numerous instan- ces of farms that have been opened in the heart of the lumbering dis- tricts, on the Menomonee, Peshitigo, Oconto, Wolf, Wisconsin, Chippe- wa and St. Croix Rivers, all of which have been attended with abundant success.

Much has been said about the " sand barrens " of Wisconsin. They lie between the upper Fox and the Wisconsin, near Fort Winnebago, and stretching off north-westerly, parallel with the Mississippi and some twenty miles from it, as far as the St. Croix. The district is generally about twenty miles in width, and some one hundred miles in length, and seems composed of what the geologists call *drift*, being the debris of sand rock, decomposed by the action of the elements in former ages, and car- ried down from the north-easterly regions by currents of water. This kind of land has been supposed entirely worthless for all purposes of agriculture. Of that part north-west of the river Wisconsin, little is known or proved as to its soils. It may be as poor as the geologists would have us believe. But it is to be remembered, that the portion of it lying between the Wisconsin and the Fox rivers, constitutes a part of the famous " Indian Land " district, which has been sought with so much avidity, during the past season, by experienced farmers from the older States.

I have, thus far, been commenting, in the main, on that portion of northern Wisconsin watered by the several rivers before mentioned, and lying below (south) of the "Region of the Lakes." The great " water shed" merits a separate consideration. As before remarked, it is about 50 miles in width, and some 150 miles in length, and will, at some future day, assume an importance in the geographical and physical features of the State now but little suspected. It is dotted over with thousands of lakes, from twenty rods to as many miles in diameter,—water clear, shores bold, filled with fine fish, abounding with water-fowl, and matchless in picturesque scenery and beautiful locations. Most of them have inlet and outlet of fine streams, while others have neither.

Let not the reader be startled with visions of eternal snows and rigors of climate; for, if correctly considered, the temperature of the Lake Region will be judged more mild than that of the next hundred miles south of it.

A mistaken notion seems to have prevailed with regard to climate, which has been too generally referred to latitude, and confounded with it; just as though the same parallels east and west necessarily always had the same degree of cold or heat. This error is now being dissipated, and climate is known to be governed by many other causes than mere latitude. Topography—geological features—go far to modify the temperature and affect the seasons; among all which, it is found that bodies of water, especially small lakes, have an influence in softening, to an extraordinary degree, a climate otherwise rigorous. Such is known to be exactly the effect in the Lake Region under consideration, and eminently so in the autumnal season.

The natives, with the sagacity of their race, made these little seas their resting places, having their villages both for summer and winter on their borders, while a large district south of them has only been used for hunting grounds. The early traveler, whether penetrating from Lake Superior, or from the lower Wisconsin and Mississippi, has always found inhabitants in the Lake Region, and been surprised to find there fields of corn, beans, potatoes, &c., which, to use the words of one of them, were "better indeed than are usually grown with all the aid of cultivation in the valley of the Ohio."—*Atwood's Geological Report, p.* 57.

Northern Wisconsin, then, as considered with regard to soil, climate, health, accessibility and markets, may be set down, in general terms, as bidding fair to become a farming country, not only "self sustaining," but producing a large surplusage for exportation. Those portions just being opened, will be the first to be settled; but the last district named, the "Region of the Lakes," will finally become the favorite.

The whole is a timbered country, prairies and openings being so few and far between, as to form exceptions to the general rule. The class of settlers, therefore, who seek the great West to find farms "already cleared to hand," will not be likely to locate in Northern Wisconsin, till they have essayed the prairies, and there made the discovery, that the bounties of Providence are nearly equally distributed over the great west; and that, what is to be had without toil, is seldom worth the possession. Our immigration will come, as it has in fact thus far, from the more Northern of the older States, the character of which, needs no eulogy from me.

Nor has so desirable a region escaped attention. The Indian title is all extinguished; the United States surveys are progressing over it with great rapidity. Men of enterprise and forecast are becoming sensible of its future destined greatness; explorations of it are being made; the eyes of capitalists are scanning it closely, evidenced in the fact that already three or four most important railways are proposed through it, to wit: one from Milwaukee *via* Portage City to La Crosse; one from Madison (a continuation of the Chicago and Beloit Railroad,) up the valley of the Wisconsin, to Ontonagon of Lake Superior; one from Fond du Lac, (a continuation of the R. R. U. V. Railroad,) up the valley of the Wolf, to Lake Superior; and one from Green Bay west to St. Paul. These roads are eminently practicable, and if constructed, would, no doubt, be remunerative on the capital invested. Indeed, from the zeal and confidence manifested by the several projectors, there is good reason to presume it not utopian to look for their early completion.

You inquire for the "prominent points in Northern Wisconsin."

But few are known as yet. Green Bay, Muckwau, Waupacca, Stevens' Point, Plover, and La Crosse, are in its southern border; though none of them, except Green Bay and Stevens' Point, are within the boundaries I have named for this part of Wisconsin. Settlements north of town 24, exists as follows: on the tributaries of Green Bay, at Oconto, Peshitigo and Menomonee rivers; on the Wolf River at Lake Shawauno; on the Wisconsin, are the Little Bull Falls village, in town 28; Big Bull Falls, (otherwise called Wausau, the seat of justice for Marathon Co.,) in town 29; besides several mill locations higher up, as at Pine

River and Virgin Falls, in town 31. I cannot give the locations with accuracy on the Black River, the Chippewa and the St. Croix, though it is known, several fine villages are thriving on those streams.

But it is hardly probable the prominent points, *in futuro*, are yet reached in this vast region; and when they shall be, they will rise in that most interesting country, the "Lake Region," and take position on the banks of the "Lake of the Desert," Lake Flambeaux," Lake Courtoreilles," and "Long Lake."

You demand to know its present wants !

What it needs first is, to be made known. 2d. To have thoroughfares opened through it. And third, and most essentially, it wants a just and liberal policy maintained towards it on the part of both the State and National Governments.

The State Agricultural Society can do much towards the first of these objects; and I trust you will not be discouraged by the feeble effort of this present paper, from soliciting treatises in future from more competent pens.

In regard to thoroughfares, I will observe, the people of Green Bay are about to petition Congress for a donation of the Public Lands, for the purpose of aiding in the construction of a railway from that town to St. Paul. Their efforts should be received with favor and helped forward by every well wisher of the North.

The proposition of the Chicago Railway Company, should be entertained in a spirit of liberality by the people of this State. Lake Superior is bound to find connexion with New York, sooner or later, *via* Chicago.

A just and liberal policy from the State and National Governments.

A part of this, so far as relates to the National Government, has been hinted at in the proposed grant of Public Lands to aid in the construction of railroads. The numberless other modes in which such questions may be raised between us and that Government, can hardly be anticipated. There is one question, however, which being already forced upon us, I beg leave to notice in this connection.

Our exertions have, hitherto, been earnestly directed towards the extinguishment of the Indian title, and the clearing of our territory of those tribes. To our petitions the General Government made prompt response, apparently in good faith. The treaty of 1848 with the

Menominee Indians, which was early ratified by the Senate, gave us the last Indian reservation; and with it an earnest that the northern part of Wisconsin would be speedly opened to the wants of settlers, and that the tribe making the cession would long before this time have left our borders. This just expectation has, however, been defeated. It is a question, pertinent to the citizen, and one which demands the profound attention of the Legislative and Executive Departments of the State, why has the law of the land—the Treaty of 1848, not been complied with, and the Menomonee Indians removed to Crow Wing River? And further, by what authority, and in pursuance of what and whose policy is it now attempted to recede twelve townships of the most valuable part of that cession back to that tribe, and to locate and re-establish them at Lake Shawauno? It is respectfully submitted for the consideration of the law makers and the Executive Authority of the State, whether, on the ratification of the treaty with the Menominee Indians of October, 1848, by the Senate of the United States, the jurisdiction of Wisconsin did not immediately become complete over all the country described in that cession? And if so, what authority, short of that of this State, can undertake to locate an Indian tribe within its borders?

Such, however, has been attempted by the Agents of the Indian Department; a tract of twelve townships of the best land in the State has been attempted to be set off near Lake Shawauno, on the Wolf river, as a home for the Menominee Indians, in violation of the rights of many of our citizens—of solemn treaty stipulation—of the jurisdiction of the State of Wisconsin—besides being all a fraud on the poor Indians themselves, who it is manifest can have no permanent resting place there; our settlements, having already, since the ratification of the treaty, preceded the pretended reservation!

Northern Wisconsin is deeply injured by this proceeding, and wishes her Government to protect itself from insult, and its citizens, whether many or few in number, from this wrong attempted to be inflicted on them in the name of the United States Indian Department.

In the foregoing, allusion is had to the policy of the National Government. I now beg leave to speak of some acts and doings of our local State Government.

It is the plain duty of every Government, to protect the weak as well as the strong; indeed, the strong being able to protect themselves, Governments are instituted, in this republican country at least, for the express purpose of protecting the weak. When they forget this, and lend their omnipotence to aid the more powerful part of the body politic, to wrong and oppress the weaker, they become nuisances, a curse instead of a blessing, and forfeit all claim to the respect of mankind.

I beg leave to look back a little in our history. On the admission of Illinois into the Union, her politicians plead to have her northern boundary placed at 42° 30' N. It was objected to this, that the Ordinance of 1787, for the government of the Territory N. W. of the River Ohio, provided, that her Northern boundary should be a line E. and W., drawn through the southerly bend of Lake Michigan. The objection was overruled, and she was admitted, with her northern boundary as she desired. The friends of Wisconsin, (for she had few inhabitants at that time within her borders) saw a violation of her rights, murmured their disapprobation, and the matter passed on, with a loss to Wisconsin of nearly eight thousand square miles, from off her southern border.

Next came the admission of Michigan, with her contest with Ohio about boundary, which all but involved the two States in bloodshed and civil war. Wisconsin was still nearly unable to be heard in the great councils of the nation; though one magnanimous voice proclaimed her rights.* But she had no representation, much less any votes. To appease the unrighteous thirst for dominion on the part of Ohio, her boundaries, as established by the Ordinance of 1787, were again violated; and as a compensation to Michigan for the robbery on her by the former State, one-fifth part of the balance of our Territory was sliced off on the N. E. and handed over to Michigan. To this outrage we submitted with impatience, and justly complained of the high handed robbery, as a ruthless disregard of its sacred obligations on the part of Congress. But submit we were compelled to, being without the power of resistance. We comforted ourselves, however, as well as we could with the reflection, that though we had lost our rights, our honor remained, and that on attaining our majority, we would take good care to make ourselves heard and respected in the National Councils.

At length the time came for our own admission to the great confederacy, which was done with our boundaries including what the rapacity of our older and stronger sisters had left us; our fair proportions, though

---

* The late Hon. John Quincy Adams.

curtailed at both extremities, were yet regarded such as might still enable us to command a respectable rank in the great republican family.

Well, we regarded the question of boundary, which had given us so much trouble and alarm, as finally put forever at rest; and we went on in earnest, to settle up and improve our fair domain, with a view to future greatness as well as immediate convenience. But in the short space of four years, before we have fairly composed our minds and become quiet of our apprehensions from without, a new project of dismemberment rises up from within; so strange and startling is the proposition, that but few of our citizens can believe it real, but are disposed to regard it as some poor hoax, got up by grave legislators in a leisure moment, during a dearth of legitimate business, or while awaiting perhaps the action of the Executive on Bills which their greater industry had presented for his consideration. But I regret to say no such charitable excuse can be plead for our Solons. Their proposition to violate the integrity of our boundaries stands on record, in the form of a grave legislative act, entitled, *"Joint Resolution in relation to the erection of the Territory of Superior."* Then follows the Resolve, that the assent of Wisconsin "is hereby given" to the dismemberment of the State! and that fully one half of it may be set off into a distinct Territorial Government, to wit: all north of Town 30, and west of Range 10 N. of the Meridian.

The question has been asked, "where did this scheme originate?"— Though difficult to trace its paternity, it is easy to say where it did not come from, and that is, it did not originate with the people.

But the assent of Wisconsin "is given," that the best half of the State may be "set off." Now it is generally supposed that boundaries constitute a part—a most essential part of our Constitution. Has our Constitution been changed—amended? If so, when, where, and how? and if not, by what authority has the Legislature undertaken to give the "assent" of the people of this State to a sundering of it in the midst?

As for the people of Northern Wisconsin, they regard the proposition with disapprobation, and concern; its consummation would be their ruin. They are told that the interest of the northern and southern portion of the State are "so diverse!" Indeed! and suppose they should be so, what State in this Union, of respectable dimensions, but has diverse interests? The argument is too insignificant to merit pursuit. The whole scheme is liable to a construction, not greatly to the honor of the originators for magnanimity of feeling, or for that just state pride which ought to swell the bosom of every man worthy of a seat in our Legislative Halls.

Among the present wants of Northern Wisconsin, this project certainiy is not one, nor can it be regarded otherwise than as antagonistical to that just and liberal policy towards us, which I have spoken of, and which, as pioneers of the wilderness, in the van of civilization, we have a right to expect and demand.

I am, Sir, with great respect,
Your obedient servant,
ALBERT G. ELLIS.

To ALBERT C. INGHAM, Esq.
    Sec. of the Wis. State Agr. Society.

FAUNA AND FLORA OF WISCONSIN.

The following Catalogues, prepared at the request of ALBERT C. INGHAM, Esq., the Secretary of the Wisconsin State Agricultural Society, include the Animals, the Fossils in the rocks, and the Plants, heretofore observed in that State. They are very far from being complete lists of our Fauna and Flora, only such species having been included as have been actually observed by me, or were communicated to me by competent persons. A few species are added on the authority of naturalists who have formerly visited this country. They embrace 62 Mammals, 287 Birds, 19 Reptiles, 14 Fishes, 90 Mollusks, 92 Fossils, and 949 Plants.

I. A. LAPHAM.

A SYSTEMATIC CATALOGUE OF THE ANIMALS OF WISCONSIN,

PREPARED FOR THE WISCONSIN STATE AGRICULTURAL SOCIETY,
BY I. A. LAPHAM, OF MILWAUKEE, 1852.

MAMMALIA.

ORDER CARNIVORA.

DIDELPHUS, Linn.
   Virginiana, Pennant. Oppossum. Green County. (Dr. Hoy.)

VESPERTILIO, Linn. The Bat.
   *Noveboracensis, Linn. New York Bat. Milwaukee.
   pruinosus, Say. Hoary Bat. Racine. †
   *subulatus, Say. Little Brown Bat. Racine.
   *noctivagans, Le Conte. Silver Haired Bat. Racine.

---

† The species marked Racine were observed near that place, and communicated to me by Dr. P. R. Hoy.

* The specimens of the species marked thus (*) are preserved in the Collection of the Nat. Hist. Association, at Madison.

Condylura, Illiger.
    *cristata, Linn. Star-nose Mole. Milwaukee.

Scalops, Cuvier.
    aquaticus, De Kay, (Sorex aquaticus, Linn.) Common Shrew Mole.
      Racine.

Sorex, Linn.
    Dekayi, Backman. Dekay's Shrew. Racine.
    Forsteri, Richardson. Forster's Shrew. Racine.
    brevicaudus, Say. Short-tailed Shrew. Milwaukee.
    Richardsonii Backman. North West Territory. (Mr. De Kay.)
    Cooperi Bachman. North West Territory. (Mr. De Kay.)

Ursus, Linn. The Bear.
    *Americanus, Pallas. Black Bear. Milwaukee.

Procyon, Storr.
    *lotor, Linn. Raccoon. Milwaukee.

Meles, Brisson.
    Labradoria, Sabine. Badger. Milwaukee.

Gulo, Storr.
    *luscus, Linn. Wolverine. North West Territory. (Mr. Say.)†

Mephitis, Cuvier.
    Americana, Desmarest. Skunk. Milwaukee.

Mustela, Cuvier.
    Canadensis, Linn. Fisher. Milwaukee and Watertown.
    *martes, Linn. Marten. North West Territory. (Mr. Say.)
    pusilla, Dekay. Little Weasel. Racine.

Putorius, Cuvier.
    *Noveboracensis. Ermine Weasle, North West Territory. (Mr. Say.)
    *vison, Linn. Mink. Milwaukee.

Lutra, Ray.
    *Canadensis, Sabine. Otter. Milwaukee.

Canis, Linn. The Dog.
    familiaris, Linn. Indian Dog.

---

† In Long's Second Expedition.

Lupus.   The Wolf.
     occidentalis, Richardson.   Common Wolf.   Milwaukee.
     *latrans (Canis latrans, Say.)   Prairie Wolf.   Racine.

Vulpes.   The Fox.
     *fulvus, Desm.   Red Fox.   Milwaukee.
     Virginianus, Dekay.   Grey Fox.   Racine.

Felis, Linn.
     concolor, Linn.   Panther.   Northern Wisconsin.   (Dr. Hoy.)

Lyncus, Gray.   Lynx.
     borealis, Temminck.   Lynx.   Milwaukee.
     *rufus, Tem.   Wild Cat.   Milwaukee.

ORDER RODENTIA.

Sciurus, Linn.   The Squirrel.
     *leucotis, Gappar.   Grey Squirrel.   Milwaukee.
     *vulpinus, Gmelin.   Fox Squirrel.   Milwaukee.
     *niger, Say.   Black Squirrel.   Milwaukee.
     *Hudsonicus, Harlan.   Red Squirrel.   Milwaukee.
     *striatus, Linn.   Striped Squirrel.   Milwaukee.

Pteromys, Illiger.
     *volucella, Harlan.   Flying Squirrel.   Milwaukee.
     sabrinus, Richardson.   Winnebago County.   (Dr. Hoy.)

Spermophilus, F. Cuvier.
     *tridecimlineatus, Mitchell.   Gopher.   Milwaukee.
     grammurus, Say.   Line-tailed Squirrel.   N. W. Territory.   (Mr. Say.)
     Parryi, Richardson.   Racine.

Arctomys, Linn.
     morax, Gmel.   Woodchuck.   Racine.

Meriones, Illiger.
     *Americanus, Barton.   Deer Mouse.   Milwaukee.

Castor, Linn.
     fiber, Linn.   Beaver.†

---

† The last Beaver killed, in the southern part of Wisconsin, was in 1819, on Sugar Creek, Walworth county, a very large one.   (S. Juneau, Esq.)

Fiber, Illiger.
*zibethicus, Linn.   Muskrat.   Milwaukee.

Hystrix, Linn.
*Hudsonius, Brisson.   Porcupine.   Lake of the Hills, Sauk Co.

Mus, Linn.
decumanus, Pallas, Brown Rat (introduced.)   Racine.
rattus, Linn. Black Rat (introduced.)   Racine.
*musculus, Linn.   Mouse.   Milwaukee.
leucopus, Richardson.   Jumping Mouse.   Racine.

Arvicola, Lacepede.
riparius, Ord.   Marsh Meadow Mouse.   Racine.
hirsutus, Emmons.   Beaver Field Mouse.   Racine.
xanthognathus, Leach.   Yellow Cheeked Meadow Mouse.   Racine.

Geomys, Richardson.
bursarius, Say.   Pouched Rat.   N. W. Territory.   (Mr. Say.)

Lepus, Linn.
*nanus, Schreber.   American Grey Rabbit.   Racine.
*Americanus, Erxleben.   Rabbit.   Milwaukee.

ORDER RUMINANTIA.

Bison, Smith.
Americanus, Gmelin.   Buffalo.†

Antilope, Smith.
Americana, Ord.   Antilope.   N. W. Territory.   (Mr. Say.)

Cervus, Brisson.
*Virginianus, Linn.   Deer.   Milwaukee.
alces, Linn.   Moose.   N. W. Territory.   (Mr. Schoolcraft.)

Elaphus.
Canadensis, Ray.   Elk.   N. W. Territory.   (Mr. Say.)

Rangifer.
tarandus, Linn.   Rein-deer.   Borders of L. Superior. (Mr. Shoolcraft.)

---

† Last seen east of Mississippi in 1832.

THE BIRDS OF WISCONSIN

Having been carefully studied by Dr. Hoy, of Racine, I have obtained his permission to insert here his "Notes," published in the 6th volume of the Proceedings of the Academy of Natural Sciences at Philadelphia, instead of the catalogue prepared by me. There are some additions and corrections made by Dr. Hoy, and I have added the common names of each species. I. A. L.

NOTES ON THE ORNITHOLOGY OF WISCONSIN.

BY P. R. HOY, M. D., OF RACINE, WISCONSIN.

With few exceptions, the facts contained in the following brief Notes, were obtained from personal observations made within fifteen miles of Racine, Wisconsin, lat. 42° 49' N.; long. 87° 40' W. This city is situated on the western shore of Lake Michigan, at the extreme southern point of the heavy timbered district, where the great prairies approach near the lake from the west, and is a remarkably favorable position for ornithological investigation. It would appear that this is a grand point, a kind of rendezvous, that birds make during their migrations. Here, within the last seven years, I have noticed 287 species of birds, about one-twentieth of all known to naturalists, many of which, considered rare in other sections, are found here in the greatest abundance. It will be seen that a striking peculiarity of the ornithological fauna of this section is, that southern birds go further north in summer, while northern species go further south in winter than they do east of the great Lakes.

[* Indicates those known to nest within the State.]

VULTURINÆ, (1 species.)

*CATHARTES AURA, Linn. Turkey Buzzard.

Found occasionally as far north as Lake Winnebago, lat. 44°. More numerous near the Mississippi River.

FALCONIDÆ, (19 species.)

*AQUILA CHRYSÆTOS, Linn. Golden Eagle.

I have a fine specimen, shot near Racine, Dec. 1853.
It is a fact worthy of note that this noble eagle, in the absence of rocky cliffs for its eyrie, does occasionally nest on *trees*. One instance occurred between Racine and Milwaukee, in 1851. The nest was fixed in the triple forks of a large oak.

Haliætus Washingtonii, Aud.   Washington Eagle.

I procured, in 1850, a living bird that had been slightly wounded, which answered to Audubon's description of this *doubtful* species.   I kept it in an ample cage upwards of two years, but before its death it underwent changes in plumage which led me to believe that, had it lived, it would have proved to be the white-headed species.   I put several species of hawks and owls into the same apartment, several of which the eagle killed and devoured without ceremony.   When a fowl was introduced, he pounced upon it, and without attempting to kill, proceeded to pluck it with the greatest unconcern, notwithstanding its piteous screams and struggles.

It is my opinion that the Bird of Washington will prove to be only an unusually marked large and fine immature white-headed eagle.   My specimen, a female, measured 7½ feet in alar extent.

*Haliætus Leucocephalus, Linn.   Bald Eagle.

Numerous throughout the State.   I have seen one of these fine birds pounce upon and capture a fish in the lake.   The eagle did not disappear wholly under the water, which led me to suspect that the fish was in some way disabled.

*Pandion Haliætus, Linn.   American Fish Hawk.

Not uncommon.

*Falco Peregrinus, Gmel.   Duck Hawk.

This noble falcon is frequently met with, although not so numerous as many other hawks.   A pair nested for several years within ten miles of this city ; constructing their nest on the top of a large red beech tree.

I have seen one of these daring hawks make a swoop into a flock of Blue-winged Teal, killing two on the spot.

*Falco Columbarius, Linn.   Pigeon Hawk.

This active little falcon is numerous, especially in spring and fall, during the migration of the smaller birds.   A few nest with us, many more in the pine forests of the northern part of the State.   Those that nest in this vicinity, regularly morning and evening, visit the lake shore, in quest of bank swallows, which they seize with great dexterity while on the wing.

Falco Æsalon, Gmel.   The Merlin.

I have met with three individuals of this small species, Nov. 15th, 1849, Dec. 25th, 1850, and Dec. 12th, 1852.

*Falco Sparverius, Linn.   Sparrow Hawk.

Common.

*Astur Atricapillus, Wilson.   American Goshawk.

This daring and powerful hawk is to be found at all seasons ; the old birds only remain during winter, the young retiring further south.   The young are so different in their plumage from the old birds, that few would suspect their identity ; they are more bold and daring, much more destructive to the poultry yards than the more sly and cautious old ones—a peculiarity not, however, confined to this species.

*Astur Cooperii, Bonap.   Cooper's Hawk.

Not uncommon.   They destroy many quails, and young grouse, which, together with poultry, constitute their principal fare. They construct their nests on the top of large trees, in the most secluded situations, and leave us at the approach of winter.

*Astur Fuscus, Gmel.   Slate colored Hawk.

Common.   Nest here about the middle of April.

Ictinea Plumbea, Gmel. Mississippi Kite.

I saw a single specimen of this southern kite on Rock River, in this State, in July 1846. It is occasionally met with on the Mississippi River.

***Nauclerus Furcatus, Linn. Swallow Tailed Hawk.**

This kite was numerous within ten miles of Racine, where they nested up to the year 1848, since which time they have abandoned this region. I have not seen one since 1850. They nested on tall elm trees about the 10th of June, and left us about the 1st of September.

***Buteo Lagopus, Wilson. Rough Legged Falcon.**

Not numerous. I have repeatedly seen this buzzard *soar* to great heights, notwithstanding the testimony of some ornithologists to the contrary. They are in the habit, while in the pursuit of mice, frogs, &c , of balancing themselves over marshy situations on the prairies. If not successful, they *sail* off to other more suitable grounds, and renew the same motion. When they espy the quarry, they dart directly upon it, in the manner of the true falcon. Where there are trees, they may adopt a different mode of hunting. My observations apply to the prairies.

***Buteo Borealis, Gmel. Red tailed Hawk.**

Common. They do not remain with us during severe winters. I have a fine albino specimen of this species. Although pure white, the irides were *yellow*. This individual was knowh to inhabit a particular district in Huron Co., Ohio, for ten years. Although I had offered a liberal reward for the capture of the " white hawk," it was several years before I succeeded in getting him.

***Buteo Vulgaris, Willoughby. (?) Common Buzzard.**

Not numerous. One of our winter residents. It is now probable this will prove a new species, and will be named Buteo Bairdii. (Hoy.)

***Buteo Pennsylvanicus, Wilson. Broad winged Buzzard.**

Common.

***Buteo Lineatus, Gmel. Winter Buzzard.**

This noisy species is extremely numerous. The great number of hawks, of this and other species, that are often seen soaring in company during fine weather, about the 20th of September, at which time they are migrating south, is almost incrédible.

***Circus Cyaneus, Linn. Marsh Harvier.**

Common. They build their nests entirely of grasses (carex,) placed on the ground in the middle of swampy marshes. Nest about the 1st of June.

STRIGINÆ, (14 species.)

Surnia Funerea, Gmel. Hawk-Owl.

A few visit us every winter. -

Surnia Nyctea, Linn. Snowy Owl.

Numerous on the prairies from November to March.

***Scops Asio, Linn. Screech Owl.**

Common. In the month of June I caught four young ones just as they were about leaving the nest. They were a deep reddish brown, in all respects similar to the old female which I shot at the time, and have preserved.

***Scops Nevia, Gmel. Mottled Owl.**

Common. I am not yet satisfied that the mottled and red owls are specifically the same.

***Bubo Virginianus, Gmel. Great Horned Owl.**

One of our most numerous species. I once put a remarkably large and fine owl of this species into the same cage with the " Washington Eagle," previously mentioned, which soon resulted in a contest. The moment a bird was given to the owl, the eagle demanded it in his usual peremptory manner, which was promptly resisted with so much spirit and determination, that for a time I was in doubt as to the result; but finally the eagle had to stand aside, and witness the owl devour the coveted morsel. After several similar contests, it was mutually settled that possession gave an undisputed right, the owl not being disposed to act on the offensive. I had a fine red-shouldered hawk in the same aviary, which the owl killed and ate the second night.

Bubo Subarcticus, Hoy. White-bellied Horned Owl.

This winter visitor I consider closely allied, yet distinct from the common horned owl, and as such it is described in the Proc. Acad. Nat. Sci. vol. vi. page 211. I have as yet examined but three specimens. The specimen in the collection of the Academy was known to carry off from one farm, in the space of a month, not less than twenty-seven individuals of various kinds of poultry, before it was shot.

Surnium Cinereum, Linn. Great Grey Owl.

Not numerous. I have a fine male specimen, shot near Racine, Jan. 4, 1848. A remarkable peculiarity of this specimen was, that the irides were *brilliant blood-red.* I saw one size carry off a duck on Lake Superior, near the mouth of Cerf river, Sept. 1st, 1845.

***Syrnium Nebulosum, Linn. Barred Owl.**

Common in the heavy timbered districts.

***Otus Vulgaris, Aud. Long-eared Owl.**

More numerous in the vicinity than any other owl. The young leave the nest about the middle of June.

***Otus Brachyotus, Linn. Short-eared Owl.**

Common on the prairies, where they nest on the ground, in the tall grass. The young are fully fledged by the second week in June.

***Nyctale Acadica, Gmel. Acadian Owl.**

Common.

Nyctale Tengmalmi, Gmel. ——

I procured a single specimen near Racine, Nov. 30th, 1850. Not uncommon on the head waters of the Wisconsin river.

Nyctale Kirtlandii, Hoy. Kirtland's Owl.

A third species of this genus, found here, and described in the Proc. Acad. Nat. Sci. vol. vi. page 210.

Only two specimens have yet been observed.

STRIX FLAMMEA, Linn. American Barn Owl.

A fine specimen of this handsome owl was obtained this spring by my friend the Rev. A. C. Barry of this city. It was shot near this city, and is the only specimen which has come under my observation.

CAPRIMULGIDÆ, (2 species.)

*ANTROSTOMUS VOCIFERUS, Wilson. Whip Poor Will.

Common. Arrives about the 1st of May, departs middle of September.

*CHORDEILES VIRGINIANUS, Briss. Night Hawk.

Numerous. They leave us by the 15th of September. On the 10th of this month, 1850, for two hours before dark, these birds formed one continuous flock, moving south. They reminded me, by their vast numbers, of passenger pigeons, more than night hawks. Next day not one was to be seen.

HIRUNDINIDÆ, (6 species.)

*PROGNE PURPUREA, Linn. Purple Martin.

Common.

*HIRUNDO AMERICANA, Wilson. Barn Swallow.

Numerous.

*HIRUNDO FULVA, Vieill. Cliff Swallow.

A few nested, for the first time, at Racine in 1852. Numerous in many parts of the State.

*HIRUNDO BICOLOR, Vieill. White-bellied Swallow.

Not a numerous species with us. Arrives from the 1st to the middle of April.

*COTYLE RIPARIA, Linn. Bank Swallow.

This numerous species perforates the banks of the lake, wherever the soil is sandy.

*CHÆTURA PELASGIA, Temm. Chimney Swallow.

Common.

HALCYONIDÆ, (1 species.)

*CERYLE ALCYON, Linn. King-fisher.

Common.

LANIADÆ, (3 species.)

*LANIUS BOREALIS, Vieill. Northern Butcher Bird.

This large shrike is most numerous during fall and winter. A few, however, spend the summer in the middle and northern parts of this State. During winter they subsist on field mice (arvicola) and small birds.

23

*Lanius Ludovicianus, Linn.? ——

I much doubt whether the north-western and southern loggerhead are the same. Our bird is *smaller* than the southern, as described in the ornithological works, the adult male measuring 8¼ to 12¼ ; female 8¼ to 10¼. The nest and eggs, too, differ materially from Bachman's description, as quoted by Nuttall, of those of the southern species. The Wisconsin bird constructs a compact nest. placed on the lower branches of a small tree. It is composed externally of small sticks and roots, filled in with strips of bark and the lint of various species of plants, and is amply lined with feathers, which almost conceal the six *spotted* eggs.

The male assists in incubation, which is completed in fourteen days.

I once shot a female just as the pair had commenced building. The male went on and completed the nest, even to the soft lining of feathers, then took his stand on the topmost branch of the same tree, and continued watching almost constantly for three days, apparently awaiting the return of his mate. At the end of that time I missed him, and supposed he had abandoned the spot; but the second day afterwards he returned with a new bride, who appeared well satisfied with the waiting nest, and commenced laying immediately.

They return to a particular tree to nest every year. This attachment is so great, that when the nest is destroyed, even after they commence incubation. they not unfrequently construct another on the same tree. Mice, young birds and large insects compose their fare.

Numerous on the border of the prairies. Arrive 1st of April ; depart in October.

Lanius Excubitoroides, Swains. ——

I shot a pair of birds of this species in March last (1853.) The female is faintly marked on the breast with pale brown undulating lines. This is undoubtedly a distinct species.

MUSCICAPIDÆ, (14 species.)

*Tyrannus Intrepidus, Vieill. King Bird.

Common.

*Tyrannus Crinitus, Linn. Great Crested King Bird.

Not so common as the preceding. Inhabits the dark swampy woods, where the harsh *squeak* of this species is frequently heard.

*Tyrannula Fusca, Gmel. Dusky Fly-catcher.

This familiar pewee is met with everywhere.

*Tyrannula Virens, Linn. Wood Pewee.

Common in the deep solitary woods.

Tyrannula Phœbe, Lath. Phœbe Bird.

I shot two individuals of this species May 10th, 1848. Probably not very rare, but impossible to distinguish it from the T. fusca without carefully comparing the two.

*Tyrannula Acadica, Gmel. Green-crested Fly-catcher.

The most numerous of the fly-catchers in Wisconsin.

Tyrannula Pusilla, Swains. ——

This species, so closely allied to the preceding, is not unfrequently met with about the 10th af May, on its passage north.

Tyrannus Cooperii, Nutt.   Olive-sided King Bird.

I have occasionally met with this bird during the latter part of May.

*Setophaga Ruticilla, Linn.   American Red Start.

Numerous.   Arrive 5th of May ; commence constructing their nests 1st of June.

*Setophaga Mitrata, Bonap.   Hooded Warbler.

Not numerous near Racine, which may be considered the northern limit of this bird's summer migration.

*Setophaga Canadensis, Linn.

This interesting species is not uncommon with us.

Setophaga Wilsonii, Bonap.

Common from the 10th to the 25th of May.

Setophaga Minuta, Wilson.

Rarely met with.   The only specimens I have, were shot 1st of June, 1850.

*Culicivora Ceruleo, Linn.   Blue-grey Gnat-catcher.

Not uncommon.   Arrives first of May.

VEREONINÆ, (6 species.)

*Vireo Flavifrons, Vieill.   Yellow-throated Greenlet.

Not uncommon.   First appearance from 10th to 15th of May.

*Vireo Solitarius, Vieill.   Solitary Greenlet.

This is by no means a rare bird in Wisconsin ; it frequents the most secluded thickets.   Arrives about 15th of May.

*Vireo Noveboracensis, Gmel.

I have noticed but few specimens of this species.   Not common.

*Vireo Gilvus, Vieill.   Warbling Greenlet.

This cheerful songster is rather common with us.

*Vireo Olivaceus, Linn.   Red-eyed Greenlet.

By far the most abundant of the birds of this genus ; its sprightly and melodious song is heard almost constantly during the summer.

*Icteria Viridis, Gmel.   Yellow-breasted Chat.

A few only are to be found in the tangled thickets during the summer months.

MERULIDÆ, (10 species.)

Mimus Polyglottus, Lath.   Common Mocking Bird.

Occasionally a straggler of this charming songster finds its way as far north as Wisconsin.   I saw one between Racine and Kenosha, July 16th, 1851, and a second near the State line, on Rock River, July 26th, 1846.

*Mimus Rufus, Linn.   Brown Thrush.

Very abundant.

*Mimus felivox, Bonap.   Cat Bird.

Common.

*Turdus Migratorius, Linn.   Robin.

Abundant.   Arrives middle of March, leaves first of November.

*Turdus Mustelinus, Gmel.   Wood Thrush.

Common. Wishing to add to my collection a pair of this species, together with their nest and eggs, I shot the female, and was about to secure the nest, when the male, which had been watching me in the vicinity, commenced singing; and, as I approached the spot he glided off still farther from the nest, all the time pouring forth the most mellow and plaintive strains I ever before heard uttered by this most melodious of songsters.   After I had been enticed to a considerable distance, he returned to the vicinity of the nest; three or four times I followed this bird in the same manner before I succeeded in shooting him.   This movement, and the effect of his tender song, so far enlisted my sympathies that I regretted exceedingly my cruelty in destroying his nest and mate.   It is common for birds to resort to various stratagems for the purpose of attracting intruders from their nests, but this is the only instance with which I am acquainted where the charms of their music were employed for this object.

Turdus Solitarius, Wilson. (?) Hermit Thrush.

Numerous during spring and fall.
Is our bird, which retires further north to breed, the same that nests in the Southern States?

*Turdus Wilsonii, Bonap.   Wilson's Thrush.

Common.   Nests 1st of June.

*Seiurus Noveboracensis, Gmel.   New York Water Thrush.

Abundant in spring and fall. A few nest in dark and gloomy swamps. Their song is sweet, a mixture between the Warbler's merry ditty, and the more mellow strains of the Thrush.

*Seiurus Aurocapillus, Wilson.   Oven Bird.

Common.

Anthus Ludovicianus, Lichst.   American Titlark.

Abundant on the prairies in spring and fall.

SYLVIADÆ, (36 species.)

Sylvicola Coronata, Wilson.   Myrtle Bird.

Numerous. The first warbler that arrives in the spring—1st of April; they all go north by the third of May; in the fall they linger with us until November.

Sylvicola Petechia, Lath.   Red-poll Warbler.

Very numerous, especially in the fall, when thousands may be seen any day on the prairies—running along the fences—flitting from stalk to stalk in the cornfield—all the time wagging their tails in the manner of the Titlark and Aqutic-Thrush, which they closely resemble in habits.

349

*Sylvicola Æstiva, Gmel.   Summer Yellow Bird.

Abundant.

Sylvicola Maculosa, Lath.   Spotted Warbler.

Numerous from the 5th to 27th of May.

Sylvicola Flavicollis, Wilson.   ——

I shot a single individual of this species near Racine, June 20th, 1848.

*Sylvicola Virens. Lath.   Black-throated Green Warbler.

Common.   A few nest with us.   The old males arrive 5th of May, young males and females about the 10th of the same month.

*Sylvicola Blackburniæ, Lath.   Blackburnian Warbler.

One of the most numerous warblers from the 5th to the 20th of May.   The old males precede the females about a week.   The first arrivals of this species, as well as all others, are in the finest plumage.   A few nest with us.

Sylvicola Kirtlandii, Baird.   Kirtland's Warbler.

I met one single individual of this recently discovered species, at Racine, May 20th, 1853.

*Sylvicola Icterocephala, Lath.   Chestnut-sided Warbler.

This beautiful little warbler is extremely abundant.   It prefers localities with a dense under-brush, especially hazel, thinly covered with trees.   In such situations it is not uncommon to hear the songs of a dozen males at the same time.   They construct a nest of blades of grass and thin strips of bark intermingled with caterpillars' web, fixed in a low bush, (generally hazel,) seldom more than two or three feet from the ground ; the eggs, 4 or 5 in number, closely resemble those of *S. æstiva*. But one brood is raised in a season—nest from the 10th to the 15th of June.   If the nest be approached when the female is in it, she will drop to the ground and hobble along with one wing dragging, uttering at the same time a *peeping* note of distress.

I once caught a young bird of this species that had just left the nest ; the parent birds, in their alarm for its safety, approached so near to me that I caught the male in my hand.   I let them both go, upon which, the joy of the old bird appearded to be greater for the escape of the young fledgling than for his own release.

Sylvicola Castanea, Wilson.

Arrives in large numbers about the 10th of May ; all gone north by the 25th.

Sylvicola Striata, Lath.   Black-poll Warbler.

Equally numerous with the preceding ; the two species arrive and depart in company.

Sylvicola Pinus, Lath.   Pine Warbler.

Not a numerous species with us.   Nest in the northern pine forests.

Sylvicola Discolor, Vieill.   Prairie Warbler

A few are occasionally seen about the middle of May.   Rare in Wisconsin.

***Sylvicola Americana, Lath. Yellow-throated Black Warbler.**

Common. The beautiful pensile nest of this bird has never, to my knowledge, been discribed. Audubon undoubtedly erred in attributing the nest described by him to this species. That presented by me to the collection of the Academy, is formed by interlacing and sewing together, with a few blades of grass, the pendant lichen (Usnea barbata) which grew upon a dead horizontal branch of an oak, fifty or sixty feet from the ground. A hole, just large enough for the bird to enter, is left in the angle immediately under the branch, which forms a complete roof for the nest; it is finished with a slight lining of hair. The whole forms a beautiful basket of moss, which is so admirably adapted to the purpose intended, so effectually concealed, so light and airy, that it would be almost impossible to suggest an improvement, and is certainly one of the most interesting specimens of ornithological architecture. The eggs, four or five in number, are white, with a band of light brown spots near the greater end; they measure 5 by $7\frac{1}{4}$ lines in diameter. The young leave the nest about the first week in July.

Sylvicola Canadensis, Linn. Spotted Canada Warbler.

Abundant from the 5th to 25th of May, and again from the 1st to 20th of October.

Sylvicola Formosa, Bonap. Kentucky Warbler.

Rare. Shot one near Racine, May 10th, 1851.

***Sylvicola Cœrulea, Wilson. Blue-grey Warbler.**

Not common. A few nest with us.

Sylvicola Maritima, Wilson. Cape May Warbler.

By no means a rare bird during the month of May. It frequents the vicinity of streams and swamps that abound with tall willows, in the tops of which this interesting warbler is commonly seen flitting about, busily searching for insects and their larvæ. It is probable that a few nest in this vicinity.

***Trichas Marylandica, Wilson. Yellow Throat.**

Common.

Trichas Agilis, Wilson.

Not uncommon. I shot a pair on the 29th of May; they had mated, and were about to nest.

Trichas Philidelphia, Wilson. Mourning Warbler.

Rarely seen. Shot one May 10th, 1851.

***Vermivora Pennsylvanica, Swain. Worm-eating Warbler.**

A few nest in this section. Rare.

***Vermivora Chrysoptera, Linn. Golden winged Warbler.**

Not uncommon. Nests with us.

Vermivora Rubricapilla, Wilson. Nashville Warbler.

Common for two weeks in May on their passage north; they return in October, at which time the male is without the chestnut crown.

***Vermivora Celata, Say. Orange-crowned Warbler.**

Not an uncommon species. Frequent the most secluded swamps, where they nest.

Vermivora Peregrina, Wilson.   Tennessee Warbler.

Some seasons, about the middle of May, this plain bird is met with in great abundance. This was particularly the case May 14th, 15th and 16th, 1849, when I could have procured any desired number; they literally thronged on the tops of the bush oaks in an adjoining grove. For the last two years I have not procured a single specimen.

*Mniotilta Varia, Vieill.   Varied Creeping Warbler.

Common.

Mniotilta Borealis, Nutt.?   ——

I have met with specimens that answered to Nuttall's description, yet I am inclined to consider it a variety of the preceding.

In order to give some idea of the abundance and great variety of the warblers which visit us, I append a list shot in the forenoon of May 5th, 1852, by Rev. A. C. Barry and myself:

6	Sylvicola Americana,	1	Sylvicola æstiva,
1	" pinus	4	" canadensis.
1	" striata,	1	" petechia.
5	" icterocephala,	4	" maritima,
4	" virens,	1	" Setophagia canadensis,
6	" Blackburnia,	2	" Vermivora rubricapilla,
5	" maculosa,	2	" Trichas marylandica,
1	" coronata,	3	" Mniotilta varia.

47

All, except three, males in unusually fine plumage, the females not having yet arrived.

We could have obtained many more of most of the species, had it been desirable.

*Troglodytes Ædon, Vieill.   House Wren.

Common. First appearance 15th of April.

*Troglodytes Hyemalis, Vieill.   Winter Wren.

Common. Nest in abundance on the shores of Lake Superior.

*Troglodytes Brevirostris Nutt-   Short-billed Wren.

A few nest in the vicinity of Racine. Not abundant.

*Troglodytes Palustris, Wilson.   Marsh Wren.

Abundant on all reedy flats.

Troglodytes Ludovicianus, Bonap.   Mocking Wren.

I met a single wren of this species, July 5th, 1852. Undoubtedly nests sparingly in the southern part of the State. Rare.

Regulus Calendula, Linn.   Ruby-crowned Kinglet.

Abundant spring and fall.

Regulus Satrapa, Lichst.   Golden-crested Kinglet.

Abundant. Arrives 1st of April, and remains until May 10th.

Sialia Wilsonii, Swains.   Blue Bird.

The first arrival of this harbinger of spring at Racine, was—

In 1846,	-	-	March	25th.
" 1847,	-	-	"	20th.
" 1848,	-	-	"	17th.
" 1849,	-	-	"	11th.
" 1850,	-	-	"	21st.
" 1851,	-	-	"	15th.
" 1852,	-	-	"	12th.

CERTHIADÆ, (5 species.)

*Certhia Americana, Bonap.   Brown Creeper.

Common throughout the year.

*Sitta Carolinensis, Linn.   White-breasted Nuthatch.

Common, remains during the winter.

*Sitta Canadensis, Linn.   Red-bellied Nuthatch.

This species does not remain with us during winter.   A few nest near Racine, a greater number in the pine regions in the northern part of the State.

*Parus Atricapillus, Linn.   Black-cap Tit.

Abundant, remain during winter.

Parus Hudsonicus, Lath.   ——

A small party of this northern species visited Racine during the unusually cold January of 1852.

AMPELIDÆ, (2 species.)

Bombycilla Garrula, Vieill.   Black-throated Waxwing.

Arrives in large parties from the first to the last of November, and leaves by the 15th April.   The first arrivals are all young birds, destitute of the yellow markings on the wing, and with less of the wax-like appendages.   These young birds generally proceed further south to winter, while the old birds, in perfect plumage, arrive later, and seldom, if ever, go further.   I never have seen an individual entirely destitute of *wax* ornaments.   The only perceptible difference between the sexes is in size, the females being slightly the larger.   In fifty specimens accurately measured the average was :

Females,	-	-	-	8 5-12—14¼
Males,	-	-	-	8 2-12—13¾

They are unsuspicious, permitting a near approach.   Their fare consists of a variety of berries, but those of the mountain ash, (*Pyrus Americana*,) appear to be preferred to all others.   They are frequently seen to eat snow as a substitute for drink.

*Bombycilla Americana, Swain.   Cedar Bird.

Common, does not remain during winter.

ALAUDINÆ, (2 species.)

*Otocoris Alpestris, Linn.   Horned Lark.

Abundant on the prairies.   A few remain during the entire winter.

*Otocoris Rufa, Aud.

Not an abundant species with us; becomes more numerous as you go west.

FRINGILLIDÆ, (33 species.)

Plectrophanes Nivalis, Linn. White Snow Bird.

Abundant from November to April.

Plectrophanes Lapponica, Linn. Lapland Snow Bird.

Met with in great abundance on the prairies, from the middle of October to the middle of May. Before they leave us in the spring they are in full song and perfect plumage. They sing in concert like blackbirds, either while on the wing or settled on fences.

Plectrophanes Smithii, Aud. ——

Occasionally met with in considerable numbers on the prairies.

*Zonotrichia Iliaca, Bonap. Fox-colored Sparrow.

Common during October and April.

*Zonotrichia Melodia, Wilson. Song Sparrow.

Common.

Zonotrichia Pennsylvanica, Lath. White-throated Sparrow.

Abundant during spring and fall.

*Zonotrichia Leucophrys, Gmel. White-crowned Sparrow.

Met with in great abundance in company with the preceding. A few nest in the vicinity.

Zonotrichia Graminea, Gmel. Bay-winged Sparrow.

Occasionally seen, but rare.

*Zonotrichia Passerina, Wilson. Yellow-winged Bunting.

Not uncommon in the reedy *slews* on the *prairies*.

*Zonotrichia Pusilla, Wilson. Field Bunting.

Not an abundant species with us.

*Zonotrichia Socialis, Wilson. Chip Bird.

Common, arrive 1st of May.

Zonotrichia Pallida, Swains. ——

Not unfrequently met with about the middle of May.

Zonotrichia Canadensis, Lath. Tree Bunting.

Very numerous autumn and spring, but few remain through the w'nter.

*Zonotrichia Savanna, Bonap. Savanna Bunting.

Common on high prairies.

*ZONOTRICHIA LINCOLNII, Aud. Blue Striped Bunting.

Not uncommon spring and fall. A few remain during summer, and nest with us.

*NIPHEA HYEMALIS, Linn. Snow Bird.

Common spring and autumn. Do not remain through the winter. Nest on Lake Superior.

*AMMODROMUS PALUSTRIS, Wilson. Swamp Finch.

Common.

LINARIA MINOR, Aud. Lesser Red-poll.

Abundant every winter.

LINARIA BOREALIS, Temm. Mealy Red-poll.

The only time I ever met with this bird was in December, 1850.

*CHRYSOMITRIS TRISTIS, Linn. Yellow Bird.

Common.

*CHRYSOMITRIS PINUS, Wilson. Pine Finch.

Abundant.

*CHONDESTES GRAMMACEA, Say. ——

Common. Frequently met with in the roads, expanding and closing their fan-like tails at every hop. One of the most agreeable singing birds. Their song is a singular combination of the Thrush, Finch, and Tohe-Bunting. They build their nest in the open field or high prairie, under the protection of a weed or small shrub; it is constructed externally of fine grass placed in a slight excavation in the ground, and finished with a lining of hair. The eggs, four, are bluish-white, marked with straggling hair streaks and serpentine lines of dark-brown, closely resembling those of the orchard Oriole, but less pointed.

*EUSPIZA AMERICANA, Gmel. Black-throated Bunting.

Not uncommon.

*SPIZA CYANEA, Wilson. Indigo Bird.

Common.

*PIPILO ERYTHROPHTHALMA, Wilson. Ground Robin.

Abundant.

*CARPODACUS PURPUREUS, Gmel. Crested Purple Finch.

Common during spring and fall. A few nest with us, many more on the shores of Lake Superior.

CORYTHUS ENUCLEATOR, Wilson. Pine Bull Finch.

Numerous during severe winters.

*LOXIA CURVIROSTRA, Linn. American Cross-bill.

Abundant in the pine forests. Large flocks occasionally visit our vicinity during fall and winter, feeding on the seed of the sunflower (*Helianthus annuus.*)

Loxia Leucoptera, Gmel.   White Winged Crossbill.

Occasionally visit us—not common.

*Pitylus Cardinalis, Linn.   Cardinal Grosbeak.

A few stragglers nest with us—rare.

*Cocoborus Ludovicianus, Wilson.   Rose-breasted Grosbeak.
Common.   Arrive 1st of May.

Cocoborus Vespertinus, Cooper.   Evening Grosbeak.

Not an uncommon bird.   During winter and spring they frequent the maple woods, feeding on the seed of the sugar maple (*Acer saccharinum*,) in quest of which they spend much time on the ground.   I have noticed this bird as late as the 15th of May.   In all probability they nest within the State.   Unsuspicious, easily approached.   Their song lacks the melody of our other species of Grosbeaks.

*Pyranga Rubra, Wilson.   Black-winged Red Bird.

Common.

STURNIDÆ, (10 species.)

*Sturnella Ludoviciana, Linn.   Meadow Lark.

Common, but does remain during winter.

*Sturnella Neglecta, Aud.

A few visit the Lake shore in early winter, we have a specimen examined by Prof Baird and pronounced to be undoubtedly this species; it was shot on the 24th of December, when the preceding were all gone.

*Yphantes Baltimore, Linn.   Golden Oriole.

Abundant.

Yphantes Spurius, Gmel.   Orchard Oriole.

Common.

*Dolichonyx Oryzivora, Linn.   Bob o'link, or Rice Bird.

Abundant.

Molothrus Pecoris, Wilson.   Cow Bunting.

Common.   I found the egg of this bird, in one instance, in the nest of the Red-winged Blackbird.

*Agelaius Xanthocephalus, Bonap.   Yellow-headed Troopial.

A few nest within fifteen miles of Racine, in an extensive marsh.   Seldom visit the lake shore.

*Agelaius Phœniceus, Linn.   Red-winged Blackbird.

Abundant every where.

*Scolecophagus Ferrugineus, Lath.   Rusty Blackbird.

Common fall and spring.   Arrive 15th of March.   A few remain during summer.

*Quiscalus Versicolor, Vieill.   Common Crow Blackbird.

Common.

CORVIDÆ, (5 species.)

*Cyanocorax Cristatus, Linn.   Blue Jay.

Common through the year.

Cyanocorax Canadensis, Linn.   Canada Jay.

Occasionally, during severe winters, visit the vicinity of Racine.

Pica Melanoleuca, Aud.   Magpie.

Occasionally a straggler visits us.   Two were shot in Caledonia, ten miles from Racine. December, 1848.   A gentleman of this city obtained one at Bailies Harbor, on Lake Michigan, November 15, 1849.

*Corvus Americanus, Aud.   Common Crow.

A singular fact in relation to the Crow is, that it never takes up its quarters within fifteen or twenty miles of Lake Michigan, within this State.   At Racine it may be considered one of the rarest birds.

*Corvus Corax, Linn.   Raven.

More numerous than the preceding.   Remain through the winter.

TROCHILDÆ, (1 species.)

*Trochilus Colubris, Linn.   Humming Bird.

Common.

PICIDÆ, (9 species.)

*Picus Pileatus, Linn.   Crested Woodpecker.

Common in heavy timber districts.

Dendrocopus Canadensis, Gmel.

Occasionally met with during winter—rare.

*Dendrocopus Villosus, Linn.

Abundant through the year.

*Dendrocopus Pubescens, Linn.   Downey Woodpecker.

Common—remain during winter.

*Dendrocopus Varius, Linn.   Yellow-bellied Woodpecker.

Common.   Leave us 1st of November, arrive 15th April.   This Woodpecker visits the orchards during September and October, to feed upon the inner bark of the peach and cherry, girdling the stems so effectually as not unfrequently to kill the trees.   I have watched them while thus engaged in my own garden, and have carefully examined, under a microscope, the contents of the stomachs of numerous specimens.

*Melanerpes Erythrocephalus, Linn.   Red-headed Woodpecker.
Common, migratory.

Apternus Arcticus, Swains.   Arctic Woodpecker.
I have specimens of this Woodpecker shot near Racine in the month of November.

*Colaptes Auratus, Linn.   Golden-winged Woodpecker.
Common.

*Centurus Carolinus, Linn.   Red-bellied Woodpecker.
Not an abundant species with us.   They remain during winter.

CUCULIDÆ, (2 species.)

*Coccyzus Americanus, Linn.   Yellow-billed Cuckoo.
Not so numerous as the following.

*Coccyzus Erythrophthalmus, Wilson.   Black-billed Cuckoo..
Abundant.

PISTTACIDÆ, (1 species.)

Conurus Carolinensis, Linn.   Paraket.
Formerly Parakets wer ecommon on the Mississippi, within this State, latterly they are seldom met with.

COLUMBIDÆ, (2 species.)

*Ectopistes migratoria, Linn.   Wild Pigeon.
Abundant.

*Ectopistes Carolinensis, Linn.   Turtle Dove.
Common.   Remain during winter.

PAVONIDÆ, (1 species.)

*Meleagris Gallopavo, Linn.   Wild Turkey.
Formerly Turkeys were common in this section, but now none are to be found.—
The last noticed near Racine was in November, 1846.   Abundant in the south-western countries.

TETRAONIDÆ, (6 species.)

*Ortyx Virginiana, Linn.   Quail.
Within a few years this Partridge has become remarkably numerous.

*Bonasa Umbellus, Linn.   Partridge.
Common in all the timber districts.

*Tetrao Canadensis Linn. Spruce Grouse.

Common on the head waters of Wolf River and vicinity of Lake Superior.

*Tetrao Cupido, Linn. Prairie Hen.

Greatly abundant. Two sportsmen, with one dog, generally bag from fifty to eighty in a day. We challenge the world for finer sporting grounds than the prairies of Wisconsin furnish during August, September and October.

*Tetrao Phasianellis. Linn. Sharp-tailed Grouse.

Formerly quite common near Racine—now seldom met with. Abundant in all the north-western counties.

*Lagopus Saliceti, Swains. Willow Grouse.

In December, 1846, two specimens were caught in a trap ten miles from Racine. Nest in the tangle of evergreen swamps of the north-western parts of the State.— Not numerous.

RALLIDÆ, (6 species.)

*Gallinula Galeata, Lich. Florida Gallinule.

Abundant as far north as Lake Winnebago, latitude 44°.

*Fulica Americana, Gmel. American Coot.

Common in all large marshes.

*Rallus Elegans, Aud. Meadow Hen.

Abundant, nest in the prairie *slews*.

*Rallus Virginianus, Linn. Mud Hen.

Common.

*Ortygometra Carolina, Linn. Sora Rail.

Greatly abundant spring and fall; a few remain during summer to nest.

*Ortygometra Noveboracensis, Lath. New York Rail.

By no means uncommon. The young of this and the preceding three species of Rail, are fully fledged by the 15th of August.

GRUIDÆ, (10 species.)

Grus Ameircana, Forster. American Crane.

A few white sand-hill cranes are occasionally seen in the western part of the State, near the Mississippi, but never approach the Lake shores, where the following species is common. It would appear that the white is a more southern species than the brown.

*Grus Canadensis, Temm. Brown Sand-hill Crane.

Found on all our large prairies. Although we have seen large flights of these birds, we never saw, or heard of, a white individual within one hundred miles of Lake Michigan. A pair has nested regularly for fifteen years in a swamp nine miles from Racine, (we have noticed them ourselves regularly for the last seven years,) and they still continue in color unchanged. The locality of this nest is in a few tussocks of grass, in the midst of an almost impenetrable swamp, the nest is composed of coarse grass, built up in a conical form eighteen inches or two feet high, so situated that when the parent bird sits upon, or rather astride of this pyramidal nest, her feet hang down on either side into the water. The old nest is regularly repaired every spring.

Tantalus Loculator, Linn.

There is a fine specimen of this southern bird in the museum of the Wisconsin State Historical Society, at Madison, which was shot near Milwaukee, Sept. 1852.

*Ardea Herodias, Linn,   Great Blue Heron.

A common species.

*Ardea Virescens, Linn.   Green Heron.

Not uncommon in the wooded swamps of the timbered districts, never met with in the prairie marshes.

*Botaurus Lentiginosus, Swains.   American Bittern,

Abundant in the marshes and *slews* of the prairies.   The young are fully fledged by the 20th July.   We have witnessed the bittern emit his peculiar call, "pump-au-gah;" the head is drawn up to the breast, the neck being much dilated, when the first syllable *pump* is uttered in a heavy low tone, the second syllable *au* is emitted with a partial extension of the neck, and the final *gah* is accompanied with a violent darting forward of the head to the full extent of his long neck.   This ludicrous performance is repeated three or four times in succession.

*Ardeola Exilis, Bonap.   Small Bittern.

Abundant on the reedy marshes, never found in the dark, shaded, woody smamps.

Egretta Leuce, Jardins.   Great White Heron.

A single individual shot near Racine, June, 1851.

*Egretta Candidissima, Gmel.   White-crested Heron.

Not an uncommon species along the borders of small lakes.   Nest in communities, on trees in Tamarack swamps.

Egretta Caerulea, Jard.   Blue Heron.

Shot one August 28th, 1848, on Root River.

CHARADRIADÆ. (6 species.)

Charadrius Marmoratus, Wagler.   Golden Plover.

Visit us in great numbers spring and fall.

Charadrius Melodius, Ord.   Piping Plover.

Occasionally met with in the fall, not numerous.

*Charadrius Vociferus, Linn.   Kill-deer Plover.

Common, arrive from 18th to 25th of March.

Charadrius Semipalmatus, Bonap.   American Ring Plover.

A few only met in May and October.   Rare.

Charadrius Helveticus, Linn.   Whistling Plover.

Not abundant, April, May and September.

Strepsilas Interpres, Linn,   Turnstone.

Common, spring and autumn.

SCOLOPACIDÆ, (27 species.)

TRINGA ALPINUS, Linn.   Black-breasted Sand Piper.
Only met with sparingly, April and October.

TRINGA SHINZII, Brehm.   Schintz's Sand Piper.
A rare species with us, spring and fall.

TRINGA PECTORALIS, Bonap.   Pectoral Sand Piper.
We have only noticed this species in autumn.

TRINGA RUFESCENS, Vieill.   Rough-breasted Sand Piper.
Quite common from September 15th to October 10th.   Never met in the spring.

TRINGA MARITIMA, Brunnich.   Purple Sand Piper.
Greatly abundant from 15th of April to 20th of May.

TRINGA MINUTA, (?) Leister.
Not common.   A few found on the borders of small lakes.

*TRINGA PUSILLA, Wils.   Wilson's Sand Piper.
Common.   Nest in the reedy marshes.

TRINGA CINEREA, Wils.   Red-breasted Sand Piper.
We have only met this bird in October.   Rare.

TRINGA SEMIPALMATA, Wils.   Semipalmated Sand Piper.
Shot several October 1st, 1850.   Rare.

TRINGA DOUGLASII, Swains.   Long-legged Sand Piper.
Shot two April 10th, 1848.   Rare.

CALIDRIS ARENARIA, Illiger,   Sanderling.
Common on the Lake Shore spring and fall.

TOTANUS SEMIPALMATUS, Lath.   Willet.
We have met this species as late as the 10th of June.   Not numerous.

*TOTANUS VOCIFERUS, Wils.   Varied Tatler.
Abundant.   Nest in all large marshes.

*TOTANUS FLAVIPES, Lath.   Yellow Legs.
Common.

*TOTANUS SOLITARIUS, Wilson.   Solitary Tatler.
Not uncommon.

*TOTANUS MACULARIUS, Wilson.   Spotted Sand Lark.
Common.

*Totanus Bartramius, Wilson.   Grey Plover.
>   Abundant.   Nest on the high rolling prairies.

Limosa Fedoa, Linn.   The Marlin.
>   Not an uncommon bird.   We saw a pair on a marshy slew near Wisconsin river, June 15th, 1848, where they were probably nesting.

Limosa Hudsonica, Lath.   Ring-tailed Marlin.
>   We shot a single bird of this well marked species November 1st, 1850.

Macrorhampus Griseus, Leach.   Dowitchee.
>   Found sparingly spring and fall.

*Scolopax Wilsonii, Temm.   Common American Snipe.
>   Common, nest abundantly with us.

*Rusticola Minor, Vieill.   American Woodcock.
>   The first woodcock noticed in this section was in 1847, since which time they have been rapidly on the increase.

Recurvirostra Americana, Linn.   American Avoset.
>   We saw a pair on a marsh near Fox river, July 26th, 1846, where they had probably nested; we also met with a small party on the Des Plaine, May, 1847.

Himantopus Nigricollis, Vieill.   The Lawyer.
>   Met a small flock of these singular birds near Racine, April, 1847.

*Numenius Longirostris, Wilson.   Long-billed Curlew.
>   Common on large thinly settled prairies.   We found them nesting in abundance on Sun Prairie, Columbia County; also within six miles of Ceresco.

*Numenius Hudsonicus, Lath.   Jack Curlew.
>   Common spring and fall.   We found a few nesting near Fox Lake, June 15, 1848.

Numenius Borealis, Lath.   Esquimaux Curlew.
>   Met with in company with the preceding in early spring and fall.   Rare.

PINNATIPEDES, (2 species.)

Phalaropus Fulicarius, Bonap.   Red Phalarope.
>   Met with a small flock first of November, 1847.   Rare.

*Lobipes Wilsonii, Jardin.   ——
>   Not an abundant species.   Prof. S. F. Baird shot one near Racine, July 15, 1853. Nests sparingly in marshes.

ANATIDÆ, (29 species.)

*Anser Canadensis, Linn.   Wild Goose.
>   Greatly abundant spring and fall, and not a few remain during the summer to nest.

Anser Hyperboreus, Gmel.   Snow Goose.

> This species is seen late in the fall in large flocks, numbering sometimes not less than two hundred.

Anser Albifrons, Bechst.   White-fronted Goose.

> Met in large numbers spring and fall.

Anser Leucopsis, Bechst.   ——

> In December, 1850, there was a single barnacle goose kept about the harbor for two weeks.

Anser Bernicla, Linn.   Brant.

> Occasionally met on the lake shore.   Rare.

Anser Hutchinsii, Rich.   Hutchinson's Goose.

> Large flocks of this species occasionally visit us in the fall; rarely seen in the spring.

Cygnus Americanus, Sharpless   American Swan.

> Visit us regularly spring and fall.

Cygnus Buccinator, Rich.   ——

> This larger swan is frequently seen, and occasionally shot in our vicinity.

*Anas Clypeata, Linn.   Shoveller.

> Not uncommon.   A few nest in the prairie siews.

*Anas Boschas, Linn.   Mallard.

> Common.

*Anas Obscurus, Gmel.   Black Duck.

> Numerous in the interior—seldom visit the lake.

Anas Strepera, Linn.   Grey Duck.

> Shot 2d March, 1848, the only specimens we ever met with.

Dafila Acuta, Linn.   Pin-tail Duck.

> Common only early in spring and late in the fall.

Mareca Americana, Gmel.   American Widgeon.

> Abundant.

*Querquedula Discors, Linn.   Blue-winged Teal.

> Very abundant.   Nest in all the large slews.

*Querquedula Carolinensis, Steph. Green-winged Teal,

> Common.

*Denronessa Sponsa, Linn.   Wood Duck.

> Common.

Fuligula Rubida, Wilson.   Ruddy Duck.
Met occasionally fall and spring.   Not abundant.

Fuligula Valisneria, Wilson.   Canvass Back.
Rarely met.   March and October.

Fuligula Ferina, Linn.   Red Head.
Not uncommon.

***Fuligula Marila, Lann.   Broad Bill.**
Common.

***Fuligula Rufitorques, Bonap.   Bastard Broad Bill.**
Common.   Nests on the borders of grassy lakes.

Clangula Vulgaris, Fleming.   Whistler.
Common on the lake in winter and early spring.

Clangula Albeola, Linn.   Buffle-headed Duck.
Common.

Clangula Histrionica, Linn.   Harlequin Duck.
One shot in Racine harbor December 15th, 1851.   Rare.

Harelda Glacialis, Linn.   Old-wife.
Common on the Lake during winter and early spring.

Mergus Merganser, Linn.   Buff-breasted Sheldrake.
Common ; remain during winter.

***Mergus Serrator, Linn.   Red-breasted Sheldrake.**
Common.   A few nest with us.

***Mergus Cucculatus, Linn.   Hooded Sheldrake.**
Abundant.   Nest on the reedy flats.

PELECANIDÆ, (2 species.)

Phalacrocorax Dilophus, Swain.   Double Crested Cormorant.
Occasionally visit our rivers and small lakes   Rare.

Pelecanus Onocrotalus, Linn.   Pelican.
About the 10th of March the pelicans arrive, and after spending a few days in the small lakes, go further north.

LARIDÆ, (11 species.)

Sterna Cayana, Lath.   Cayenne Tern.
Rarely visits us.

*STERNA NIGRA, Linn. Black Tern.
>Abundant about lakelets and marshes. Never found on Lake Michigan.

*STERNA ANGLICA, Montague. Marsh Tern.
>We have but seldom met the marsh tern in this vicinity.

*STERNA HIRUNDO, Linn. Common Tern.
>Abundant. Nest on a small rocky island in the northern part of Lake Michigan

LARUS FRANKLINII, Rich. ——
>Visit us only in severe winters. Rare.

LARUS BONAPARTII, Rich. Bonaparte's Gull.
>Associate with the common Tern fall and spring, in great numbers.

LARUS TRIDACTYLUS, Linn. Three-toed Gull.
>Met on the lake, Nov., 1853.

LARUS SABINI, Sabine.
>Saw two in company with the preceding.

LARUS GLAUCUS, Brunnich. ——
>Another rare winter visitor.

*LARUI AGENTATUS, Brunnich. Winter Gull.
>The common Gull of the lakes. Nest on a rocky island in company with the common Tern.

LARUS ZONORHYNCHUS, Rich. Common American Gull.
>A rare species with us. Mr. Samuel Sircomb has a specimen shot at Milwaukee.

COLYMBIDÆ, (6 species.)

*COLYMBUS GLACIALIS, Linn. Great Loon or Diver.
>Common.

COLYMBUS SEPTENTRIONALIS, Linn. Red-throated Loon.
>Not uncommon during winter.

PODICEPS RUBRICOLLIS, Lath. Red-necked Grebe.
>Only found in winter. Rare.

PODICEPS CORNUTUS, Linn. Horned Grebe.
>Common spring and fall.

*PODICEPS CAROLINENSIS. Dipper.
>Common. Nest in marshes.

*PODICEPS CRISTATUS, Lath. Crested Grebe.
>A common species; nests on the margin of small lakes.

REPTILES.

ORDER TESTUDINATA.

CHELONURA, Fleming.
  *serpentina, Say.   Snapping Turtle.

EMYS, Brongniart.
  *picta, Gmelin.   Painted Tortoise.   Milwaukee.

CISTUDA, Fleming.
  Blandingii, Holbrook.   Blanding's Tortoise.   Prairies of Wisconsin.
                                                      (Holbrook.)

ORDER OPHIDIA.

CROTALUS, Linn.
  *durissus, Linn.   Banded Rattle Snake.   Grant Co.

CROTALOPHORUS, Gray.
  tergeminus, Holb.   Massasauga.   Racine.

EUTAINIA, Baird & Girard.
  *sirtalis, B. & G. (coluber sirtalis, Linn.   Tropidonotus tænia, De Kay.)
                                            Striped Snake.   Milwaukee.
  radix, B. & G.   Racine.

NERODIA, Baird & Girard.
  *sipedon, B. & G.   Watersnake.   Milwaukee.

SCOTOPHIS, Baird & Girard.
  vulpinus, B. & G.   Racine.

OPHIBOLUS, Baird & Girard.
  eximius, B. & G. (Coluber eximius, De Kay.) Milk Snake. Milwaukee.

BASCANION, Baird & Girard.
  *constrictor, B. & G.   Black Snake.   Milwaukee.

CHLOROSOMA, Wagl.
  *vernalis, B. & G.   (Coluber vernalis, De Kay.) Green Snake. Racine.

DIADOPHIS, Baird & Girard.
  punctatus, B. & G. (Coluber punctatus, Linn.) Ring-necked Snake.
                                                        Milwaukee.
STORERIA, Baird & Girard.
  Dekayi, B. & G. (Tropidonotus Dekayi, Holb.) Racine.
  occipito-maculata, B. & G.   Racine.

ORDER AMPHIBIA.

RANA, Linneus.
*palustris, Le Conte. Marsh Frog. Milwaukee.
*sylvatica, Le Conte. Wood Frog. Milwaukee.
halcinda, Kalm. Shad Frog. Milwaukee.

SALAMANDRA, Brongniart.
*subviolacea, Barton, Violet Salamander.

MENOBRANCHUS, Harlan.
*lateralis, Say. Banded Proteus. Milwaukee River.

FISHES.

ORDER SPINE-RAYED.

PERCA, Linneus.
flavescens, Mitchell. Yellow Perch. Milwaukee.

POMOTIS, Cuvier.
vulgaris, Cuv, Sunfish. Milwaukee.

CORVINA, Cuvier.
oscula, Lesueur. Sheephead. Milwaukee.

ORDER ABDOMINAL.

PIMELODUS, Cuvier.
catus, Linn. Cat-fish. Milwaukee.

CATOSTOMUS, Lesueur.
aureolus, Les. Mullet. Milwaukee.

LEUCISCUS, Klein.
atronasus, Mitchell. Minnow. Milwaukee.

ESOX, Cuvier.
estor, Lesueur. Muskallonge. Milwaukee.
reticulatus, Lesueur. Pickerel. Milwaukee.

SALMO, Linneus.
fontinalis, Mitchell. Brook Trout. Northern Wisconsin.
amethystus, Mitchell. Lake Trout. Milwaukee.

CORREGONUS, Cuvier.
albus, Lesueur. White Fish. Milwaukee.

AMIA, Linn.
    amia calva. Dog Fish. Milwaukee.

LEPISOSTEUS, Lacepede.
    osseus, Linn. Gar Fish. Rock River.

CARTILAGINOUS.

ACIPENSER, Linn.
    rubicundus, Lesueur. Sturgeon. Milwaukee. L. Winnebago.

MOLLUSCA.

ORDER GASTEROPODA.

ANCYLUS, Muller.
    rivularis, Say. Milwaukee and Rock Rivers.
    diaphanus, Hald. Milwaukee River.

VITRINIA, Draparnaud.
    pellucida, Drap. N. W. Territory, (Mr. Say.)

HELIX, Linn.
    albolabris, Say. Milwaukee. Menasha. Two Rivers. Sheboygan.
    alternata, Say. Milwaukee. Two Rivers. Sheboygan.
    arborea, Say. Milwaukee. Madison. Two Rivers. Sheboygan.
    chersina, Say. N. W. Territory, (Mr. Say.)
    concava, Say. N. W. Territory, (Binney.)
    fraterna, Say. Manitowoc.
    hirsuta, Say. Milwaukee. Sheboygan.
    labyrinthica, Say. Milwaukee. Two Rivers. Sheboygan.
    lineata, Say. Milwaukee. Two Rivers. Sheboygan.
    ligera, Say. N. W. Territory, (Mr. Say.)
    monodon, Racket. Milwaukee. Manitowoc. Sheboygan.
    multilineata, Say. Milwaukee. Madison.
    perspectiva, Say. Milwaukee. Madison. Two Rivers.
    profunda, Say. Milwaukee. Manitowoc. Sheboygan.
    porcina, Say. N. W. Territory, (Mr. Say.)

PUPA, Lamarck.
    armifera, Say. Milwaukee.
    corticaria, Say. Milwaukee.
    ovata, Say. [P. Modesta, Immatine.] Milwaukee.

SUCCINEA, Draparnaud.
    obliqua, Say. Milwaukee. Two Rivers.
    avara, Say. N. W. Territory, (Mr. Say.)

Bulimus, Bruguieres.

    lubricus, Brug. Milwaukee. Madison. Sheboygan.
    harpa, Pfeiffer. (Helix harpa, Say.) N. W. Territory, (Mr. Say,)

Planorbis, Lamarck.

    armigerus, Say. Milwaukee River. Muskego Lake.
    bicarinatus, Say. Milwaukee River. Four Lakes. Sheboygan.
    campanulatus, Say. Milwaukee River, Rock, Fourth, and Muskego Lakes.
                                Sheboygan.
    corpulentus, Say. N. W. Territory, (Mr. Say.)
    deflectus, Say. Milwaukee and Twin Rivers. Rock, Fourth and Mus-
                                  kego Lakes.
    exacutus, Say. Milwaukee River.
    parvus, Say. Milwaukee and Manitowoc Rivers. Muskego Lake.
    trivolvis, Say. Milwaukee and Twin Rivers. Muskego Lake.

Limnea, Lamarck.

    caperata, Say. (L. umbilicata Adams.) Milwaukee and Manitowoc
                                Rivers. Muskego Lake.
    catascopinn, Say. N. W. Territory, (Mr. Say.)
    columella, Say. (L. Macrostoma, Say.) Milwaukee.
    emarginata, Say. Four Lakes.
    fragilis, Linn. (L. Elodes, Say.) Milwaukee, Manitowoc and She-
             boygan Rivers. Rock, Muskego and the Four Lakes.
    gracilis, Say. Milwaukee River.
    jugularis, Say. (L. stagnalis, Kirtland.) Milwaukee, Sheboygan and
             Twin Rivers. Muskego, Rock, Horricon and Fourth Lakes.
    megasoma, Say. N. W. Territory, (Mr. Say.)
    umbrosa, Say. N. W. Territory, (Mr. Say.)

Physa, Draparnaud.

    heterostropha, Say. Milwaukee, Sheboygan and Twin Rivers. Rock L.
    elongata, Say. Milwaukee and Manitowoc Rivers.

Paludina, Lamarck.

    decisa, Say. Milwaukee, Sheboygan and Rock Rivers.
    isogona, Say. Rock River.
    subglobosa, Say. N. W. Territory, (Mr. Say.)

Amnicola, Gould & Haldeman.

    limosa, Say. N. W. Territory, (Mr. Say.)
    lustrica, Say. Four Lakes.

Melania, Lamarck.

    depygis, Say. Milwaukee and Sheboygan Rivers.
    elevata, Say. Milwaukee and Manitowoc Rivers.

Valvata, Muller.

    tricarinata, Say. Milwaukee River. Rock and Four Lakes.
    sincera, Say. Milwaukee River. Four Lakes.

ORDER ACEPHALA.

Unio, Bruguieres.

bullatus, Raf.* [U. pustulosus, Lea.] Rock River. Wisconsin River at Prairie du Sac.

cardium, Raf. [U. ventricosus, Barnes.] Mississippi at Prairie du Chien. Wisconsin River, (Mr. Barnes.)

cariosus, Say. Silver Lake. Fourth Lake.

clava, Lam. [U. mytiloides, Raf.] Rock River. Wisconsin at Prairie du Sac.

dilatus, Raf. [U. gibbosus, Bar.] Manitowoc. Rock and Wisconsin Rivers.

fasciolaris, Raf. [U. phaseolus, Hildreth, U. mucronatus Bar.] Wisconsin River, (Mr. Barnes.)

fasciatus, Raf. [U. crassus, U. ellipticus, and U. carinatus, Barnes.] At Prairie du Chien, (Mr. Say.) Wisconsin and Neenah Rivers, (Mr. Barnes.) Rock River.

luteolus, Lam. [U. siliquoideus, and U. inflatus, Bar.] Milwaukee and Rock Rivers. Pike Creek. Wisconsin River, (Mr. Barnes.)

metanervus, Raf. [U. nodosus, Bar.] Wisconsin River at Prairie du Sac.

nervosus, Raf. [U. zigzag, and U. donaciformis, Lea.] Wisconsin River at Prairie du Sac.

olivarius, Raf. [U. ellipsis, Lea.] Wisconsin River at Prairie du Sac.

parvus, Barnes. Neenah (Fox) River, (Mr. Barnes.)

prasinus, Conrad. [U. Schoolcraftensis, Lea.] Neenah River, (Mr. Lea.)

radiatus, Lam. Wisconsin River. (Mr. Barnes.)

rectus, Lam. [U. prœlongus, Barnes.] Rock River. Wisconsin River at Prairie du Sac. Neenah River, (Mr. Barnes.)

reflexus, Raf. [U. cornulus Bar.] Neenah River, (Mr. Barnes.)

subrostratus, Raf. [U. iris, Lea.] Milwaukee and Root Rivers.

trigonus, Lea. Milwaukee, Sheboygan and Rock Rivers.

truncatus, Raf. [U. elegans, Lea.] Wisconsin River at Prairie du Sac.

tuberculatus, Raf. [U. verrucosos, Bar.] Wisconsin River, (Mr. Barnes) Rock River.

undatus, Bar. Neenah River, (Mr. Barnes,) Wisconsin at Pr. du Sac.

verrucosus, Raf. [U. tuberculatus, Bar.] Neenah River, (Mr. Barnes) Wis. River at Prairie du Sac.

Metaptera, Rafinesque, Agassiz.

alata, Conrad, [Unio alatus, Say.] Milwaukee, Rock, Neenah, Wisconsin and Sheboygan Rivers.

fragilis, Con. [U. fragilis, Say. U. gracilis, Bar.] Sheboygan and Rock Rivers. Wisconsin River at Prairie du Sac.

Plectomerus, Conrad.

(Proc. Ac. Nat. Sc., vol. VI, p. 260.)

plicatus, Con. [Unio plicatus Bar.] Rock River. Wisconsin River at Prairie du Sac.

---

*I follow Mr. T. A. Conrad's Synopsis of the Family of Naiades, recently published, (Proc. of the Acad. of Nat. Sciences, vol. VI, p. 243,) for reasons there given, which appear to be conclusive.

Complanaria, Swainson.

    complanata, Con. [Alasmodonta complanata, Bar.] Wisconsin and Neenah Rivers, (Mr. Barnes.) Sheboygan and Rock Rivers.
    costata Con. [Alas costata Raf. A. rugosa Bar.] Wisconsin and Neenah Rivers, (Mr. Barnes.) Milwaukee and Rock Rivers.
    compressa, Con. [Symphynota compressa Lea.] Milwaukee River.

Alasmodonta, Say.

    marginata, Say. Milwaukee, Sheboygan and Rock Rivers.
    leptodon Raf. [Unio planus, Bar. U. tenuissima, Lea.] Wisconsin River, (Mr. Barnes.)

Strophites, Rafinesque.

    calceolus, Con. [Unio calceolus, Lea.] Milwaukee, Root and Rock Rivers. Rock Lake.
    edentulus, Con. [Alas edentula, Say. Anodonta edentula, De Kay.] Milwaukee and Rock Rivers.

Anodonta, Bruguieres.

    cataracta, Say. [A. fluviatilis, Lea.] Milwaukee River. Fourth Lake.
    declivis, Con. [A. plana, Lea.] Milwaukee River.
    ferussaciana, Lea. Milwaukee, Sheboygan and Rock Rivers. Pike Creek. Oconomowoc and Silver Lakes.
    imbecilis, Say. [A. incerta, Lea.] Milwaukee River.

Cyclas, Lamarck.

    similis, Say. Common.

PALÆONTOLOGY OF WISCONSIN,

Being an enumeration of the Fossil Organic Remains discovered in the rocks of that State. By I. A. Lapham.

This list of fossils will be of use in making a geological survey of the State. As an instance of the importance of the careful study of these interesting relics of a "former world," I may mention that this list shows conclusively that the rocks of this State belong to the Silurian period, and are therefore much older than the Coal Formation. We hence infer, with confidence, that no coal beds of workable value, will ever be found in Wisconsin; and all the money heretofore expended, or that will hereafter be expended by ignorant persons, in search of this valuable product, is only a useless waste of capital.

PLANTÆ.

Scolithus, Hall, Palæont. of N. Y., vol. I. p. 2.

    linearis, Hall, Pal. N. Y., vol. I. p. 2. One mile north of Lyons, Sauk Co , also near Madison.

PALÆOPHYCUS, Hall, Pal. of N. Y., I. p. 7.
  tubularis, Hall. Janesville.

PHYTOPSIS, Hall, Pal. of N. Y., I. p. 38.
  tubulosum, Hall. Mineral Point.

BUTHOTREPHIS, Hall, Pal. N. Y., I. p. 8.
  succulens, Hall, Pal. N. Y., I. p. 62. Mineral Point.

ZOOPHYTA.

STREPELASMA, Hall, Pal. of N. Y., I. p. 17.
  expansa, Hall. Doty's Island, Lake Winnebago.
  corniculum, Hall, Pal. N. Y., I. p. 69. Blue Mounds, Beetown, Exeter,
                                        Beloit, Emerald Grove.
  parvula, Hall, Pal. N. Y., I. p. 71. Mineral Point, Fairwater, Doty's I.

STICTOPORA, Hall, Pal. N. Y., I. p. 73.
  ramosa, Hall, Pal. N. Y., I. p. 51. Mineral Point.

CHÆTETES.
  lycoperdon, Hall, Pal. N. Y., I. p. 64 (Favosites lycoperdon, Say.) Bee-
                                        town, Mineral Point, Beloit.
RECEPTACULITES.
  a species resembling R. Neptuni, De France. (Coscinopora sulcata?
    Owen.) Mineral Point, Blue Mounds, Wyota, Emerald Grove,
    Whitewater, Fort Atkinson, Watertown.

GRAPTOLITHUS.
  Hallianus, Prout, Am. Jour. of Science, Vol. XI., p. 187. Osceola Mills,
                                        St. Croix River.
DIPLOPHYLLUM, Hall. Pal. N. Y., II. p. 115.
  cæspitosum, Hall. Milwaukee.

FAVOSITES.
  Niagarensis, Hall, Pal. N. Y., II. p. 125. Milwaukee.
  favosa, Goldfus. (F. striata, Say.) Milwaukee, Bailey's Harbor, Door Co.

CATENIPORA. The Chain Coral.
  agglomerata, Hall, Pal. N. Y., II p. 129. Milwaukee.
  gracilis, Hall, Fost. & Wh. Report, p. 212. Eastern shore of Green Bay.

HELIOLITES, Hall, Pal. N. Y., II. p. 130.
  pyriformis, Hall. Milwaukee.
  macrostylus, Hall, Pal. N. Y., II. p. 135. Milwaukee.

STROMATOPORA, Goldfus.
  concentrica, Goldfus. Hall, Pal. N. Y., II. p. 136. Milwaukee.

CRINOIDEA.

SCHIZOCRINUS, Hall, Pal. N. Y., I. p. 81.
nodosus, Hall.   Mineral Point, Beetown, Blue Mounds, Doty's Island.

EUCALYPTOCRINUS.
decorus, Hall, Pal. N. Y., II. p. 207, (Hypanthocrinities, Philips.) Racine.

CARYOCRINUS, Say.
ornatus, Say.   Hall, Pal. N. Y. II., p. 216.   Milwaukee, Racine.

BRACHIOPODA.

LINGULA.
prima, Conrad.   Hall, Pal. N. Y., I. p. 3.   Falls of the St. Croix.
antiqua, Hall, Pal. N. Y., I. p. 3.   Falls of the St. Croix.
obtusa, Hall, Pal. N. Y., I. p. 98.   Hazel Green, Little Butte des Morts.

LEPTÆNA.
alternata, Conrad.   Hall, Pal. N. Y., I. p. 102.   Mineral Point, Hazel
Green, Sturgeon Bay.
deltoidea, Conrad.   Hall, Pal. I., p. 106.   Blue Mounds, Mineral Point
Newark and Beloit, Rock Co.
sericea, Sowerby, Hall, Pal. N. Y., I. p. 110.   Platte Mounds, Mineral
Point, Blue Mounds, Doty's Island.
filitexta, Hall, Pal. N. Y., I. p. 111.   Mineral Point.
planumbona, Hall, Pal. N. Y., I. p. 112.   Mineral Point.
deflecta, Conrad.   Hall, Pal. N. Y., I. p. 113.   Mineral Point.
recta, Conrad.   Hall, Pal. N. Y., I. p. 113.   Patch Grove.
planoconvexa, Hall, Pal. N. Y., I. p. 114.
depressa, Delman.   Hall, Pal. N. Y., II. p. 257.   Milwaukee.
subplana, Conrad.   Hall, Pal. N. Y., II. p. 259.   Milwaukee.

ORTHIS.
testudinaria, Delman.   Hall, Pal. N. Y., I. p. 117.   Mineral Point, Platte
Mounds, Blue Mounds, Beloit, Sturgeon Bay.
subæquata, Conrad.   Proc. Ac. Nat. Sc. I. p. 333.   Hall, Pal. N. Y., I.
p. 118.   Mineral Point.
bella-rugosa, Conrad.   Hall, Pal. N. Y., I. p. 118.   Mineral Point.
disparilis, Conrad.   Hall, Pal. N. Y., I. p. 119.   Mineral Point.
perveta, Conrad.   Hall, Pal. N. Y., I. p. 120.   Mineral Point.
tricenaria, Conrad.   Hall, Pal. N. Y., I. p. 121.   Mineral Point, Blue
Mounds, Exeter, Newark, Rock Co.
pectinella, Conrad.   Hall, Pal. N. Y., I. p. 123.   Mineral Point.
occidentalis.   Hall, Pal. N. Y., I. p. 127.   Sturgeon Bay.
hybrida, Sowerby.   Hall, Pal. N. Y., II. p. 253.   Milwaukee.

Spirifer.

    lynx, Von Buch. Hall, Pal. N. Y., I. p. 133. Mineral Point, Doty's Island, Sturgeon Bay.
    Niagarensis, Conrad. Hall, Pal. N. Y., II. p. 264. Milwaukee.

Atrypa.

    increbescens. Hall, Pal. N. Y., I. p. 146. Mineral Point, Newark.
    reticularis, Dalman. Hall, Pal. N. Y., II. p. 72. (Anomia reticularis, Linn.) Milwaukee.
    nitida, Hall, Pal. N. Y., II. p. 268. Milwaukee.
    obtusicplicata, Hall, Pal. N. Y., II. p. 279. Milwaukee.

Pentamerus.

    Oblongus, Murchison. Hall, Pal. N. Y., II. p. 79. Milwaukee, Blue Mounds, Bailey's Harbor, Door Co., Sturgeon Bay.
    occidentalis, Hall, Pal. N. Y., II. p. 341. On the peninsula between Green Bay and Lake Michigan. Mr. Hall.

ACEPHALA.

Nucula.

    levata, Hall, Pall. N. Y., I. p. 150. Platte Mounds, Hazel Green.

Tellinomya, Hall, Pal. N. Y., I. p. 151.

    nasuta, Hall. Mineral Point.

Amboychia, Hall, Pal. N. Y., I. p. 163.

    Amygdalina, Hall, Pal. N. Y., I. p. 165. Mineral Point, Beetown, Beloit.

Modiolopsis.

    curta, Conrad. Hall, Pal. N. Y., I. p. 297. Mineral Point.
    ovatus, Hall, Pal. N, Y., II. p. 101. Milwaukee.

Pyrenomœus, Hall, Pal. N. Y., II. p. 87.

    cuneatus, Hall. Milwaukee.

GASTEROPODA.

Maclurea, Laseuer.

    magna, Laseuer, Jour. Acad. Nat. Sc., vol. I., p. 313, pl. 13, figs. 1, 2, 3. Hall, Pal. N. Y., I. p. 26, pl. 5, fig. 1. Newark, Rock county.

Pleurotomaria.

    Umbilicata, Hall, Pal. N. Y., I. p. 43. Beloit, Fairwater, Doty's Island,
    lenticularis, Conrad. Hall, Pal. N. Y., I. p. 172. Mineral Point, Exeter, Newark, Doty's Island, West side of Green Bay.
    subconica, Hall, Pal. N. Y., I. p. 174. Mineral Point, Newark, Rock Co.

MURCHISONIA.

tricarinata, Hall, Pal. N. Y., I. p. 178.   Mineral Point.
bellicincta, Hall, Pal. N. Y., I. p. 179.   Beetown, Blue Mounds, Exeter, Emerald Grove.
major, Hall, in Foster & Whitney's Report, p. 209.   Western Shore of Green Bay.
gracilis, Hall, Pal. N. Y., I. p. 181.   Mineral Point, Exeter, Newark, Doty's Island.

SUBULITES, Conrad.   Hall, Pal. N. Y., I. p. 182.

elongata, Conrad.   Hall.   Mineral Point, Beloit, Patch Grove.

BELLEROPHON.

bilobatus, Sowerby.   Hall, Pal. N. Y., I. p. 184.   Mineral Point, Doty's Island.

BUCANIA.

expansa, Hall, Pal. N. Y., I. p. 186.   Newark, Rock Co.

CYRTOLITES.

compressus, Conrad.   Hall, Pal. N. Y., I. p. 188.   Mineral Point.

CYCLONEMA, Hall, Pal. N. Y., II. p. 87.

bilix (Pleurotomaria? bilix, Hall, Pal. N. Y., I. p. 305.)   Near Prairie du Chien.

PLATYOSTOMA, Conrad.   Hall, Pal. N. Y., II. p. 286.

Niagarensis, Hall, Pal. N. Y., II. p. 287.   Milwaukee.
hemispherica, Hall, Pal. N. Y., II. p. 288.

CEPHALOPODA.

ORTHOCERAS.

multicameratum, Conrad.   Hall, Pal. N. Y., I. p. 45.   Beetown, Mineral Point.
bilineatum, Hall, Pal. N. Y., I. p. 199.   Hazel Green.
anellum, Conrad.   Pro. Ac. Nat. Sc. I., p. 334—Hall, Pal. N. Y., I. p. 202.   Hazel Green, Mineral Point.
undulatum, Hisinger.   (O. annulatum, Sowerby.)   Hall, Pal. N. Y., I. p. 293.   Milwaukee.

LITUITES.

undatus, Conrad, in Hall, Pal. N. Y. I., p. 52.   Beloit.
convolvulans ?   Hisinger.   Hall, Pal. N. Y., I. p. 53.   Newark, Rock County.

GONIOCERAS, Hall, Pal. N. Y., I. p. 54.

anceps, Hall.   Mineral Point.

CYRTOCERAS.

macrostonum, Hall, Pal. N. Y., I. p. 194.   Hazel Green,

Oncoceros. Hall, Pal. N. Y., I. p. 196.
    constrictum, Hall. Newark, Beloit, Fairwater.

CRUSTACEA.

Asaphus, Brongniart.
    Barrandi, Hall, in Fost. & Whitney's Report, p. 212. Near Platteville.
    extans, Hall, Pal. N. Y., I. p. 228. Patch Grove, Mineral Point.

Cytherina.
    fabulites, Conrad. Proc. Acad. Nat. Sc. Vol. I., p. 332. Beloit, Mineral
                                         Point.

Illænus.
    crassicauda, Dalman. Hall, Pal. N. Y. I. p. 229. Mineral Point, Hazel
                                         Green.

Isotelus, De Kay.
    gigas, De Kay. Hall, Pal. N. Y., I. p. 231. Mineral Point.

Ceraurus, Green.
    pleurexanthemus, Green. Hall, Pal. N. Y., I. p. 242. Beloit, Mineral
                                  Point, Patch Grove.
    insignis, Beyrich. Hall, Pal. N. Y., II. p. 300.

Phacops.
    Dalmani, Portlock. (P. callicephalus. Hall, Pal. N. Y., I. p. 247.)
                                         Patch Grove.

Bumastis.
    barriensis, Murchison. Hall, Pal. N. Y., II. p. 302. Burlington Racine,
                Co.; Milwaukee; Hartford, Washington Co.

Calymene, Brongniart.
    Blumenbchii, Brongniart. Hall, Pal. N. Y., II. p. 307. Milwaukee.
                                         Racine.

PLANTS OF WISCONSIN.

The vicinity of the " Great Lakes," Superior and Michigan; the elevated plateau between Lake Superior and the Mississippi River ; the "pineries;" the heavily timbered land ; the "oak openings," and the prairies, may each be considered as distinct botanical districts, within the State, affording plants peculiar to themselves, and giving great richness and variety to our flora.

Mr. Thomas Nuttall was the first botanist, so far as I can learn, who visited Wisconsin. He passed Green Bay, by the Portage of the Neenah

(Fox) and Wisconsin rivers, to Prairie du Chien, and thence down the Mississippi, as early as about the year 1813. In his very valuable "Genera of North American Plants," published in 1818, he makes frequent reference to localities in this State, and has described thirteen new species first discovered by him in these regions.

The next notice of our plants was published in 1821, in Sillman's Journal,* by Prof. D. B. Douglass and Dr. John Torrey; being "a notice of the plants collected by Prof. Douglass, in an expedition under Governor Cass, during the summer of 1820, around the Great Lakes and upper waters of the Mississippi." One hundred and ten plants are enumerated, many of them from within the limits of this State; and three are indicated as new species.

In 1823, Major Long, with a party of scientific gentlemen, under the direction of the Secretary of War, traversed the North West Territory (as Wisconsin was then called); but unfortunately the botanist was detained, and did not join the Expedition. We have, consequently, only an account of a few plants gathered by the late lamented Thomas Say, naturalist to the Expedition; these were examined by Lewis de Schweinitz, an accomplished botanist of Pennsylvania, and a list of them published in the Narrative of the Expedition.

The next and last published notice of our plants is in Schoolcraft's "Narrative of an Expedition through the Upper Mississippi to Itasca Lake, the actual source of that river, in 1832." This Expedition was accompanied by the late Dr. Douglass Houghton, whose premature death in Lake Superior, while performing his arduous duties of State Geologist of Michigan, is sincerely regretted, not only by all who knew him, but by all the friends of science. The list of plants collected by him in this Expedition, numbers two hundred and forty-seven, of which eight were new and undescribed.

Numerous prepared specimens of Wisconsin plants, have, within the last few years, been distributed among the botanists of our own and other countries; and the critical notices kindly returned by them, have been of much assistance in making this enumeration. It embraces one hundred and thirty-six of the natural orders or families, four hundred and fifty-nine genera, and nine hundred and forty-nine species—all found within thirty miles of the city of Milwaukee, unless other localities are mentioned.

*Vol. IV. p 56.

RANUNCULACEÆ. THE CROWFOOT FAMILY.

ATRAGENE, Linn.
    Americana, Sims. Head of Lake St. Croix. Dr. Parry.

CLEMATIS, Linn. Virgin's Bower.
    Virginiana, Linn. Common Virgin's Bower.

PULSATILLA, Tourn. Pasque-flower.
    patens, Mill. Anemone patens, Linn.

ANEMONE, Linn. Wind-flower.
    nemorosa, Linn. Low Wind-flower.
    Virginiana, Linn. Tall Anemone.
    multifida, DC. Shore of Lake Superior. Dr. Z. Pitcher.
    Pennsylvanica, Linn.

HEPATICA, Dillenius. Liver-leaf.
    triloba, Chaix. Round-leaved Hepatica.
    acutiloba, DC. Sharped-leaved Hepatica.

THALICTRUM, Linn. Meadow Rue.
    anemonoides, Michx.
    diocum, Linn.
    Cornuti, Linn. Meadow Rue.

RANUNCULUS, Linn. Crowfoot.
    aquatilis, Linn. White-Water Crowfoot.
    Purshii, Richards. Yellow-Water Crowfoot.
    rhomboideus, Goldie.
    abortivus, Linn. Small-flowered Crowfoot.
    recurvatus, Poir.
    Pennsylvanicus, Linn. Bristly Crowfoot.
    fascicularis, Muhl.
    repens, Linn. Creeping Crowfoot.
    Marylandicus, Poir. Hairy Crowfoot.
    acris, Linn. Buttercups.

ISOPYRUM, Linn. Enemion, Raf.
    biternatum, Torrey & Gray.

CALTHA, Linn. Marsh Marigold.
    palustris, Linn. Cowslip.

COPTIS, Salisb. Goldthread.
    trifolia, Salisb.

AQUILEGIA, Linn. Columbine.
    Canadensis, Linn. Wild Columbine.

25

DELPHINIUM, Linn.   Larkspur.
    azureum, Michx.   Upper Mississippi, Dr. Houghton.

HYDRASTIS, Linn.   Orange-root.
    Canadensis, Linn.

ACTÆA, Linn.   Cohosh.
    rubra, Bigel.   Red Cohosh.
    alba, Bigel.   White Cohosh.

MENISPERMACEÆ.   THE MOONSEED FAMILY.

MENISPERMUM, Linn.   Moonseed.
    Canadense, Linn.

BERBERIDACEÆ.   THE BARBERRY FAMILY.

LEONTICE, Linn.   Caulophyllum, Michx.
    thalictroides, Linn.   Blue Cohash.

JEFFERSONIA, Barton.   Twin-leaf.
    diphylla, Pers.

PODOPHYLLUM, Linn.   May Apple.
    peltatum, Linn.

CABOMBACEÆ.   THE WATER-SHIELD FAMILY.

BRASENIA, Schreber.   Hydropeltis, Michx.
    peltata, Pursh.   Water-shield.

NELUMBIACEÆ.   THE NELUMBO FAMILY.

NELUMBIUM, Juss.   Sacred Bean.
    luteum, Willd.   Upper Mississippi, Dr. Houghton.

NYMPHÆACÆ.   THE WATER-LILY FAMILY.

NYMPHÆA, Tourn.
    odorata, Ait.   White Water-Lily.

NUPHAR, Smith.
    advena, Ait.   Yellow Water-Lily.

SARRACENIACEÆ.   THE PITCHER PLANT FAMILY.

SARRACENIA, Linn.
    purpurea, Linn.   Side-Saddle Flower.

PAPAVERACEÆ. The Poppy Family.

SANGUINARIA. Dill. Bloodroot.
  Canadensis, Linn.

FUMARIACEÆ. The Fumitory Family.

DICENTRA, Bork.
  cucullaria, DC. Breeches Flower.
  Canadensis, DC. Squirrel Corn.

CORYDALIS, Linn.
  aurea, Willd. Upper Mississippi, Dr. Houghton.
  glauca, Pursh. Blue Mounds.

CRUCIFERÆ.

NASTURTIUM, R. Brown.
  palustre, DC.
  natans, DC.

CARDAMINE, Linn.
  rhomboidea, DC. Spring Cress.
  hirsuta, Linn.
  pratensis, Linn.

DENTARIA, Linn. Toothwort.
  laciniata, Muhl.

ARABIS, Linn.
  petræa, Lam. Shore of Lake Superior, Dr. Z. Pitcher.
  lyratta, Linn.
  hirsuta, Scop.
  lævigata, DC.
  Canadensis, Linn. Sicklepod.

TURRITIS, Dill. Tower Mustard.
  glabra, Linn. Shore of Lake Superior, Dr. Z. Pitcher.
  stricta, Graham. Lake Superior, Dr. A. Gray.
  brachycarpa, Torr. & Gray. Shore of Lake Superior, Dr. Z. Pitcher.

ERYSIMUM, Linn. Treacle Mustard.
  cheiranthoides, Linn. Lake Superior, Dr. Houghton.

SISYMBRIUM, Linn. Hedge Mustard.
  canescens, Nutt.

SINAPIS, Tourn. Mustard.
  arvensis, Linn.
  nigra, Linn.

DRABA, Linn. Whitlow Grass.
>Caroliniana, Walt. Near Waukesha, Mr. G. W. Cornwall.

CAMELINA, Crantz. False Flax.
>sativa, Crantz.

LEPIDIUM, Linn. Pepperwort.
>Virginicum, Linn.

CAPSELLA, Vent. Shepherd's Purse.
>Bursa-pastoris, Mænch.

CAKILE, Tourn. Sea-Rocket.
>Americana, Nutt.

CAPPARIDACEÆ. THE CAPER FAMILY.

POLANSIA, Raf.
>graveolens, Raf. Near Beloit, Mr. T. McEl Henry.

VIOLACEÆ. THE VIOLET FAMILY.

VIOLA, Linn. Violet.
>blanda, Willd.
>sagittata, Ait. Marquette County, Mr. John Townley.
>cucullata, Ait. Blue Violet.
>pedata, Linn.
>Muhlenbergii, Torrey.
>pubescens, Ait. Yellow Violet.

CISTACEÆ. THE ROCK-ROSE FAMILY.

HELIANTHEMUM, Trum. Rock Rose.
>Canadense, Michx.

HUDSONIA, Linn.
>tomentosa, Nutt. Lake Superior, Dr. Houghton.

LECHEA, Linn. Pinweed.
>minor, Lam. Upper Mississippi, Dr. Houghton.

DROSERACEÆ. THE SUNDEW FAMILY.

DROSERA, Linn. Sundew.
>rotundifolia, Linn.
>longifolia, Linn. L. Sup. to Upper Miss., Dr. Houghton.
>linearis, Goldie. La Pointe, L. Superior, Dr. Houghton.

PARNASSIA, Tourn.
    Carollniana, Michx.
    palustris, Linn. South shore of L. Superior, Dr. Z. Pitcher.

HYPERICACEÆ. THE ST. JOHN'S-WORT FAMILY.

HYPERICUM, Linn. St. John's-Wort.
    pyramidatum, Ait. Waukesha, Mr. G. H. Cornwall.
    Canadensis, Linn. Lake Superior, Dr. Houghton.
    prolificum, Linn. Marquette Co., Mr. J. Townley.

ELODEA, Adans.
    Virginica, Nutt.

CARYOPHYLLACEÆ. THE PINK FAMILY.

§ 1. *Sileneæ.*

SAPONARIA, Linn. Soapwort.
    Vaccaria, Linn. Cow-herb.

SILENE, Linn. Catchfly.
    stellata, Ait. Starry Campion.
    antirrhina, Linn.
    noctiflora, Linn.

LYCHNIS, Tourn. Cockle.
    githago, Lam.

§ 2. *Alsineæ.*

ARENARIA, Linn. Sandwort.
    stricta, Michx.
    serpyllifolia, Linn. Waukesha, Mr. G. H. Cornwall.
    lateriflora, Linn.

STELLARIA, Linn. Chickweed.
    media, Smith.
    longifolia, Muhl. Stichwort.

CERASTIUM, Linn.
    vulgatum, Linn. Beloit, Mr. T. McEl Henry.
    viscosum, Linn. Mouse-Ear.

§ 3. *Illecebreæ.*

SPERGULA, Linn. Spurrey.
    arvensis, Linn. Wauwatosa, Mr. M. Spears.

ANYCHIA, Michx.
    dichotoma, Michx. Blue Mounds, I. A. Lapham.

§ 4. *Mullugineæ.*

Mollugo, Linn.
    verticilatta, Linn. Carpet Weed.

PORTULACACEÆ. The Purslain Family.

Portulaca, Tourn.
    oleracea, Linn. Purslain.

Talinum, Adans.
    teretifolium, Pursh. Falls of the St. Croix, Dr. Houghton.

Claytonia, Linn. Spring Beauty.
    Virginica, Linn.

MALVACEÆ. The Mallow Family.

Malva, Linn. Mallow.
    triangulata, Leavenworth. M. Houghtonii, Torr. & Gray. Upper Mississippi, Dr. Houghton.
    rotundifolia, Linn. Dwarf Mallow.

TILIACEÆ. The Linden Family.

Tilia, Linn.
    Americana, Linn. Basswood.

LINACEÆ. The Flax Family.

Linum, Linn.
    rigidum, Pursh.

GERANIACEÆ.

Geranium, Linn. Cranesbill.
    maculatum, Linn.
    Carolinianum, Linn.
    Robertianum, Linn. L. Superior to Upper Mississippi, Dr. Houghton.

OXALIDACEÆ. The Wood Sorrel Family.

Oxalis, Linn.
    violacea, Linn. Rock Prairie and near Beloit.
    stricta, Linn.

BALSAMINACEÆ. The Balsam Family.

Impatiens, Linn. Jewel Weed.
    pallida, Nutt.
    fulva Nutt.

LIMNANTHACEÆ.

Flœrkea, Willd.   False Mermaid.
  proserpinacoides, Willd.

ZANTHOXYLACEÆ.   The Prickly Ash Family.

Zanthoxylum, Linn.
  Americanum, Mill.   Prickly Ash.

Ptelea, Linn.
  trifoliata, Linn.

ANACARDIACEÆ.   The Sumach Family.

Rhus, Linn.   Sumach.
  typhina, Linn.
  glabra, Linn.
  venenata, DC.   Poison Sumach.
  Toxicodendron, Linn.   Poison Oak.

ACERACEÆ.   The Maple Family.

Acer, Linn.   Maple.
  spicatum, Lam.   Mountain Maple.
  saccharinum, Wang.   Sugar Maple.
  rubrum, Linn.   Red Maple.

Negundo, Mœnch.   Box Maple.
  aceroides, Mœnch.   Rock River and Sugar River, I. A. L.

CELASTRACEÆ.   The Spindle-Tree Family.

Staphylea, Linn.   Bladder-Nut.
  trifolia, Linn.   Beloit.   Mr. T. McEl Henry.

Celastrus, Linn.   Bittersweet.
  scandens, Linn.

Euonymus, Tourn.   Spindle-Tree.
  atropurpureus, Jacq.   Here called Wahoo.

RAMNACEÆ.   The Buckthorn Family.

Rhamnus, Linn.   Buckthorn.
  alnifolius, L'Her.

Ceanothus, Linn.   New Jersey Tea.
  Americanus, Linn.
  ovalis, Bigel.   Beloit, Mr. T. McEl Henry.

VITACEÆ. The Vine Family.

Vitis, Linn. Grape Vine.
    æstivalis, Michx. Summer Grape.
    ripari, Michx. Frost Grape.

Ampelopsis, Michx. Virginia Creeper.
    quinquefolia, Michx.

POLYGALACEÆ. The Milkwort Family.

Polygala, Tourn. Milkwort.
    incarnata, Linn. Beloit, Mr. T. McEl Henry.
    sanguinea, Linn. P. purpurea, Nutt.
    crusiata, Linn.
    verticiliata, Linn.
    Senega, Linn. Seneca Snake-Root.
    polygama, Walt. Waukesha, Mr. G. H. Cornwall.
    paucifolia, Willd.

LEGUMINOSÆ. The Pea Family.

Vicia, Tourn. Vetch.
    Cracca, Linn. Tufted Vetch.
    Americana, Muhl.
    Caroliniana, Waltr. Marquette Co., Mr. J. Townly.

Lathyrus, Linn.
    maritimus, Bigel.
    venosus, Muhl.
    ochroleucus, Hook.
    palustris, Linn.

Apios, Boerh.
    tuberosa, Mœnch. Indian Potato.

Amphicarpæa, Ell.
    monoica, Nut. Wild Bean.

Desmodium, DC. Tick Trefoil.
    nudiflorum, DC. St. Croix, Dr. Parry.
    acuminatum, DC.
    Canadense, DC.

Lespedeza, Michx. Bush Clover.
    violacea, Pers. var. divergens, Beloit, Mr. T. McEl Henry. Var.
    capitata, Michx. [sessiliflora Waukesha, I. A. Lapham.

Astragalus, Linn. Milk Vetch.
    Canadensis, Linn.

PHACA, Linn.   Bladder Vetch.
neglecta, Torr. & Gray.

TEPHROSIA, Pers.   Hoary Pea.
Virginiana, Pers.

AMORPHA, Linn.
fructicosa, Linn.   Beloit, Mr. T. McEl Henry.
canescens, Nutt.   Lead Plant.

DALEA, Linn.
laxiflora, Pursh.   Near Prairie du Chien, Mr. Nuttall.

PETALOSTEMON, Michx.
violaceum, Michx.
candidum, Michx.
villosum, Nutt.   St. Croix, Dr. Parry.

TRIFOLIUM, Linn.   Clover.
pratense, Linn.   Red Clover.
repens, Linn.   White Clover.

LUPINIS, Tourn.   Lupine.
perennis, Linn.   Wild Lupine.

BAPTISIA, Vent.   False Indigo.
australis, R. Brown.   B. cærulea, Nutt.   On the Neenah River, Dr.
leucantha, Torr. & Gray.   B. alba, Hooker.   [Houghton.
leucophæa, Nutt.

CASSIA, **LINN**.   Senna.
Chamæcrista, Linn.   Beloit, Mr. T. McEl Henry.

GLEDITSCHIA, Linn.   Honey Locust.
triacanthus, Linn.   Beloit, Dr. S. P. Lathrop.

ROSACEÆ.   THE ROSE FAMILY.

§ 1. *Amygdaleæ.*

PRUNUS. Tourn.   Plum.
American, Marsh.   Yellow Plum.

CERASUS, Tourn.   Cherry.
pumilla, Michx.   Sand Cherry.   Lakes Michigan and Superior, Dr.
Pennsylvanica, Loisel.   Bird Cherry.   [Houghton.
Virginiana, DC.   Choke Cherry.
serotina, DC.   Wild Black Cherry.

§ 2. *Rosaceæ.*

SPIRÆA, Linn.   Meadowsweet.

opulifolia, Linn.   Nine-Bark.
salicifolia, Linn.   Meadowsweet.
tomentosa, Linn.   Hardhack.   Upper Wisconsin River.   I. A. L.

AGRIMONIA, Tourn.   Agrimony.

Eupatoria, Linn.

GEUM, Linn.   Avens.

Virginianum, Linn.   White Avens.
macrophyllum, Willd.   Lake Superior, Dr. Z. Pitcher.
strictum, Ait.
rivale, Linn.   Water or Purple Avens.
triflorum, Pursh.

WALDSTEINIA, Willd.

fragaroides, Tratt.   Dividing ridge between the St. Croix and Boi
[Brule Rivers, Dr. Parry

POTENTILLA, Linn.

Norvegica, Linn.
Canadensis, Linn.   Fivefinger.
paradoxa, Nutt.   Near St. Croix River, Dr. Parry.
arguta, Pursh.
anserina, Linn.   Silver-Leaf.
fruticosa, Linn.
tridentata, Ait.   Lake Superior, Dr. Houghton.

COMARUM, Linn.   Marsh Fivefinger.

palustre, Linn.

FRAGARIA, Tourn.   Strawberry.

Virginiana, Ehrh.
vesca, Linn.

RUBUS, Linn.   Bramble.

Nutkanus, Mocino.   Head of Lake Superior, Dr. Houghton.
odoratus, Linn.   Lake Superior, Dr. Parry.
canadensis, Linn. Low Blackberry. Marquette Co., Mr. J. Townle
triflorus, Richards.
strigosus, Michx.   Red Raspberry.
occidentalis, Linn.   Black Raspberry.
villosus, Ait.   Blackberry.
hispidus, Linn.   Lake Superior, Dr. Houghton.

ROSA, Tourn.   Rose.

lucida, Ehrh.   Wild Rose.
blanda, Ait.   Lake Superior, Dr. Houghton.

§ 3. *Pomeæ.*

CRATÆGUS, Linn.

    coccinea, Linn. White Thorn.
    punctata, Jacq.

PYRUS, Linn. Apple.

    coronaria, Linn. Crab-Apple.
    arbutafolia, Linn. Choke Berry.
    Americana DC. Mountain Ash. Brookfield, Waukesha Co., Mr.
                                       [G. H. Cornwall.

AMELANCHIER, Medic. June-Berry.

    Canadensis, Torr. & Gray.

MELASTOMACEÆ.

RHEXIA, Linn. Deer Grass.

    Virginica, Linn. Mauvaise River, C. Whttlesey.

LYTHRACEÆ. THE LOOSESTRIFE FAMILY.

LYTHRUM, Linn. Loosestrife.

    alatum, Pursh.

DECODON, Gmel.

    verticillatum, Ell. Fish Trap Rapids, Upper St. Croix, Dr. Parry.

ONAGRACEÆ. THE EVENING PRIMROSE FAMILY.

§ 1. *Onagraceæ.*

EPILOBIUM, Linn.

    angustifolium, Linn. Willow Herb.
    coloratum, Muhl.
    palustre, Linn.

ŒNOTHERA, Linn. Evening Primrose.

    biennis, Linn.
    rhombipetala, Nutt. Near the St. Croix, Dr. Parry.
    chrysantha, Michx. Waukesha, Mr. G. H. Cornwall.

GAURA, Linn.

    biennis, Linn. Beloit.

LUDWIGIA, Linn.

    palustris, Ell. Water Purslain.

CIRCÆA, Tourn. Enchanter's Nightshade.

    Lutetiana, Linn.
    alpina, Linn.

§ 2. *Halorageæ.*

MYRIOPHYLLUM, Vaill.   Water Milfoil.
 spicatum, Linn.
 verticillatum, Linn.

HIPPURIS, Linn.   Mares-Tail.
 vulgaris, Linn.

CACTACEÆ.   THE CACTUS FAMILY.

OPUNTIA, Tourn.   Prickley Pear.
 vulgaris, Mill.   Falls of the St. Croix, Dr. Parry.

GROSSULACEÆ.   THE GOOSEBERRY FAMILY.

RIBES, Linn.
 Cynosbati, Linn.   Prickly Gooseberry.
 hirtellum, Michx.   Smooth Gooseberry.
 rotundifolium, Michx.   Swamp Gooseberry.
 floridum, Linn.   Wild Black Currant.
 rubrum, Linn.   Wild Red Currant.

CUCURBITACEÆ.   THE CUCUMBER FAMILY.

ECHINOCYSTIS, Torr. & Gray.
 lobata, Torr. & Gray.   Wild Cucumber.

CRASSULACEÆ.   THE HOUSE-LEEK FAMILY.

PENTHORUM, Gronov.
 sedoides, Linn.   Stonecrop.

SAXIFRAGACEÆ.   THE SAXIFRAGE FAMILY.

SAXIFRAGA, Linn.   Saxifrage.
 Aizoon, Jacq.   Lake Superior, Dr. Z. Pitcher.
 Virginiensis, Michx.   Lake Superior, Dr. Houghton.
 Pennsylvanica, Linn.

HEUCHERA, Linn.   Alum Root.
 Americana, Linn.

MITELLA, Tourn.   Mitrewort.
 diphylla, Linn.   Currant Leaf.
 nuda, Linn.

CHRYSOSPLENIUM, Tour.   Golden Saxifrage.
 Americanum, Schw. Lake Superior to Up. Mississippi, Dr. Houghton

HAMAMELACEÆ. The Witch Hazel Family.

HAMAMELIS, Linn. Witch Hazel.
Virginica, Linn.

UMBELLIERÆ.

HYDROCOTYLE, Tourn.
Americana, Linn. Falls of the St. Croix, Dr. Parry.

SANICULA, Tourn.
Marylandica, Linn. Sanicle.

ERYNGIUM, Tourn.
aquaticum, Linn. Rattlesnake-Master.

POLYTÆNIA, DC.
Nuttallii, DC.

HERACLEUM, Linn. Cow Parsnip.
lanatum, Michx.

PASTINACA, Tourn.
sativa, Linn. Wild Parsnip. Poisonous.

ARCHANGELICA, Hoffm.
atropurpurea, Hoffm.

CONIOSELINUM, Fischer.
Canadense, Torr. & Gray.

ZIZIA, Koch. Alexanders.
cordata, Koch.
aurea, Koch.
integerrima, DC.

BUPLEURUM, Tourn.
rotundifolium, Linn. Introduced.

CICUTA, Linn.
maculata, Linn.
bulbifera, Linn.

SIUM, Linn. Water Parsnip.
latifolium, Linn.

CRYPTOTÆNIA, DC.
Canadense, DC. Honewort.

Osmorhiza, Raf.   Sweet Cicely.

    longistylis, DC.
    brevistylis, DC.   St. Croix, Dr. Parry.

Conium, Linn.   Poison Hemlock.

    maculatum, Linn.   Green Bay.

Erigenia, Nutt.

    bulbosa, Nutt.

ARALIACEÆ.   The Spikenard Family.

Aralia, Linn.

    racemosa, Linn.   Spikenard.
    nudicaulis, Linn.   Wild Sarsaparilla.
    hispida, Michx.   Upper Wisconsin River, I. A. L.

Panax, Linn.

    quinquefolium, Linn.   Ginseng.
    trifolium, Linn.   Ground-Nut.

CORNACEÆ.   The Dogwood Family.

Cornus, Tourn.

    alternifolia, Linn.   Yellow-Twigged Dogwood.
    circinata, L'Her.   L. Superior to Upper Miss., Dr. Houghton.
    sericea, Linn.
    stolonifera, Michx.   Red-Twigged Dogwood.
    paniculata, L'Her.
    Canadensis, Linn.   Pudding Berry.

CAPRIFOLIACEÆ.   The Honeysuckle Family.

Linnæa, Gronov.   Twin-Flower.

    borealis, Gronov.

Symphoricarpus, Dill.

    occidentalis, R. Brown.   Wolf-Berry.
    racemosus, Michx.   Snowberry.

Lonicera, Linn.   Honeysuckle.

    sempervirens, Ait.   Lake Superior, Dr. Houghton.
    flava, Sims.   Yellow Honeysuckle.
    parviflora, Lam.
    hirsuta, Eaton.   L. Superior to Upper Miss., Dr. Houghton.
    ciliata, Muhl.
    cærulea, Linn.
    oblongifolia, Muhl.

DIERVILLA, Tourn.
  trifida, Mœnch.

TRIOSTEUM, Linn. Horse-Gentian.
  perfoliatum, Linn.

SAMBUCUS, Linn. Elder.
  Canadensis, Linn.
  pubens, Michx. Lake Maria, Marquette Co., I. A. L.

VIBURNUM, Linn.
  Lentago, Linn.
  dentatum, Linn. Arrowwood.
  acerifolium, Linn.
  Opulus, Linn. V. oxycoccus, Pursh. High Cranberry.

RUBIACEÆ. THE MADDER FAMILY.

§ 1. *Stellatæ.*

GALIUM, Linn.
  Aparine, Linn. Goose-Grass.
  asprellum, Michx.
  trifidum, Linn.
  triflorum, Michx.
  lanceolatum, Torr. Wild Liquorice.
  boreale, Linn.

§ 2. *Cinchoneæ.*

CEPHALANTHUS, Linn. Button Bush.
  occidentalis, Linn.

MITCHELLA, Linn. Partridge-Berry.
  repens, Linn.

HEDYOTIS, Linn.
  ciliolata, Torr.
  longifolia, Hook. St. Louis River, Dr. Houghton.

VALERIANACEÆ. THE VALERIAN FAMILY.

VALERIANA, Tourn. Valerian.
  edulis, Nutt. V. ciliata, Torr. & Gray.

FEDIA, Gærtn.
  Fagopyrum, Torr. & Gray.

COMPOSITÆ.

VERNONIA, Schreb.

    Noveboracensis, Willd.   Upper Mississippi, Dr. Houghton.
    fasciculata, Michx.   Beloit, Mr. T. McEl Henry.

LIATRIS, Schreb.

    cylindracea, Michx.
    scariosa, Willd.
    spicata, Willd.

KUHNIA, Linn.

    eupatorioides, Linn.

EUPATORIUM, Tourn.

    purpureum, Linn.
    perfoliatum, Linn.   Thoroughwort.
    ageratoides, Linn.

ADENOCAULON, Hook.

    bicolor, Hook.   Lake Superior, Dr. Z. Pitcher.

ASTER, Linn. Starwort.

    macrophyllus, Linn.
    sericeus, Vent.
    concolor, Linn.   Neenah River, Dr. Houghton.
    lævis, Linn.
    azureus, Lindl.
    Shortii, Boott.
    cordifolius, Linn.   St. Croix, Dr. Parry.
    sagittifolius, Willd.
    multiflorus, Ait.
    miser, Linn.
    tenuifoius, Linn.   Upper Mississippi, Dr. Houghton.
    præaltus, Poir.
    carneus, Nees.
    laxifolius, Nees.
    puniceus, Linn.
    prenanthoides, Muhl.
    oblongifolius, Nutt.   Upper Mississippi, Torr. & Gray.
    Novæ-Angliæ, Linn.
    ptarmicoides, Torr. & Gray.

ERIGERON, Linn. Fleabane.

    Canadense, Linn.
    bellidifolium, Muhl.   Roberts's Plantain.
    Philadelphicum, Linn.
    glabellum, Nutt.   St. Croix River, Dr. Houghton.
    strigosum, Muhl.

DIPLOPAPPUS, Cass.

    linariifolius, Hook.   Beloit, Mr. T. McEl Henry.
    umbellatus, Torr. & Gray.

SOLIDAGO, Linn.   Golden Rod.

    bicolor, Linn.   Falls of St. Croix, Dr. Parry.
    latifolia, Linn.
    stricta, Ait. St. Croix, Dr. Parry.
    speciosa, Nutt.
    Virga-aurea, Linn.   Lake Superior, Dr. Houghton.
    rigida, Linn.
    Ohioensis, Riddell.
    Riddellii, Frank.
    neglecta, Torr. & Gray.
    patula, Muhl.
    arguta, Ait.
    altissima, Linn.
    ulmifolia, Muhl.
    nemoralis, Ait.
    Canadensis, Linn.
    lanceolata, Linn.

CHRYSOPSIS, Nutt.

    villosa, Nutt. Upper Mississippi, Dr. Houghton.

INERLA, Linn. Elecampane.

    heleniurn, Linn. Introduced.

POLYMNIA, Linn.

    Canadensis, Linn.

SILPHIUM, Linn.

    ternatum, Linn. Waukesha, Mr. G. H. Cornwall.
    laciniatum, Linn. Compass Plant.
    terebinthinaceum, Linn. Prairie Dock.
    trifoliatum, Linn. Waukesha, Mr. G. H. Cornwall.
    integrifolium, Michx.
    perfoliatum, Linn. Waukesha, Mr. G. H. Cornwall.

PARTHENIUM, Linn.
    integrifolium, Linn.

AMBROSIA, Tourn.
    trifida, Linn.
    artemisiæfolia, Linn.

XANTHIUM, Tourn.
    strumarium, Linn.

HELIOPSIS, Pers. Oxeye.
    lævis, Pers.

Echinacea, Mœnch.
angustifolia, DC.   Beloit, Mr. T. McEl Henry.

Rudbeckia, Linn.
laciniata, Linn.
hirta, Linn.

Lepachys, Raf.
pinnata, Torr. & Gray.

Helianthus, Linn.   Sunflower.
rigidus, Pers.
occidentalis, Riddell.
giganteus, Linn.
decapetalus, Linn.
strumosus, Linn.

Coreopsis, Linn.
trichosperma, Michx.
palmata, Nutt.
lanceolata, Linn.   Two Rivers, Manitowoc Co.

Bidens, Linn.   Beggar-Ticks.
frondosa, Linn.
cernua, Linn.   .
chrysanthemoides, Michx.
Beckii, Torr.   St. Croix River, Dr. Houghton.

Helenium, Linn.
autumnale, Linn.

Maruta, Cass.   May-Weed.
Cotula, DC.

Achillea, Linn.   Yarrow.
Millefolium, Linn.

Tanacetum, Linn.   Tansey.
Huronensis, Nntt.   Lake Superior, Dr. Houghton.

Artemisia, Linn.   Wormwood.
Canadensis, Michx.
Ludoviciana, Nutt.
biennis, Willd.

Gnaphalium, Linn.
decurrens, Ives.   Lake Superior, Dr. Parry.
polycephalum, Michx.
uliginosum, Linn.   Elkhorn, Walworth Co.

ANTENNARIA, Gærtn.
> dioica, Gærtn.
> plantaginifolia, Hook.

ERECHTHITES, Raf.   Fireweed.
> hieracifolia, Raf.   Falls of St. Croix, Dr. Parry.

CACALIA, Linn.   Indian Plantain.
> suaveolens, Linn.
> reniformis, Muhl.
> atriplicifolia, Linn.
> tuberosa, Nutt.

SENECIO, Linn.
> vulgaris Linn.
> aureus, Linn.   Ragwort.
> tomentosus, Michx.   Beloit, Mr. T. McEl Henry.

CERSIUM, Tourn.
> lanceolatum, Scop.   Thistle, introduced.
> Pitcheri, Torr. & Gray.   Two Rivers, Manitowoc Co.
> Virginianum, Michx.
> muticum, Michx.
> pumilum, Spreng.
> arvense, Scop.   Canada Thistle, introduced.

LAPPA, Tourn.
> major, Gærtn.   Burdock.

CYNTHIA, Don.
> Virginica, Don.

HIERACIUM, Tourn.   Hawkweed.
> Canadense, Michx.
> scabrum, Michx.
> Gronovii, Linn.
> longipilum, Torrey.   Blue Mounds.

NABALUS, Cass.
> albus, Hook.
> racemosus, Hook.

TROXIMON, Nutt.
> cuspidatum, Pursh.

TARAXACUM, Haller.
> Dens-leonis, Desf.   Dandelion.

LACTUCA, Tourn.
> elongata, Muhl.

Sonchus, Linn.   Sow-Thistle.
    oleraceus, Linn.

LOBELIACEÆ.   The Lobella Family.

Lobella, Linn.
    cardinalis, Linn.   Cardinal Flower.
    siphilitica, Linn.   Blue Lobelia.
    inflata, Linn.   Indian Tobacco.
    spicata, Lam.
    Kalmii, Linn.

CAMPANULACEÆ.   The Bell-Flower Family.

Campanula, Tourn.
    rotundifolia, Linn.   Hair-Bell.
    aparinoides, Pursh.
    Americana, Linn.

Specuiaria, Heist.
    perfoliata, A. DC. Marquette Co., Mr. J. Townley.

ERICACEÆ.   The Heath Family.

§ 1. *Vacciniew.*

Gaylussacia, H. B. & K.   Huckleberry.
    resincsa, Torr. & Gray.   Black Huckleberry.

Vaccinium, Linn.
    macrocarpon, Ait.   Cranberry.
    Pennsylvanicum, Lam.   Blue Huckleberry.
    Canadense, Kalm.   Blueberry.   Falls of St. Croix, Dr. Parry.

Chiogenes, Salisb.
    hispidula, Torr. & Gray.

§ 2. *Ericinew.*

Arctostaphylos, Adans.
    Uva-ursi, Spreng.   Bear-Berry.

Gaultheria, Kalm.
    procumbens, Linn.   Winter-Green.

Epigæa, Linn.
    repens, Linn.   Mayflower.   Lemonwier River.

Andromeda, Linn.
    polifolia, Linn.
    calyculta, Linn.

Kalmia, Linn. Laurel.
>glauca, Ait. Lake Superior, Prof. Douglass.

Ledum, Linn. Labrador Tea.
>latifolium, Ait. Dells of the Wisconsin River.

§ 3. *Pyroleæ.*

Pyrola, Linn.
>rotundifolia, Linn.
>asarifolia, Michx.
>elliptica, Nutt. Shin-Leaf.
>secunda Linn.

Moneses, Salisb. Pyrola, Linn.
>uniflora. Mauvaise River, Dr. Houghton.

Chimaphila, Pursh.
>umbellata, Nutt. Prince's Pine.

§ 4. *Monotropeæ.*

Hypopitys, Dill.
>lanuginosa, Nutt. Lake Superior, D.. Parry.

Monotropa, Gronov. Indian Pipe.
>uniflora, Linn.

AQUIFOLIACEÆ. The Holly Family.

Prinos, Linn. Winter-Berry.
>verticillatus, Linn.

Nemopanthes, Raf. Lemonwier River.
>Canadensis, DC.

PLANTAGINACEÆ. The Plantain Family.

Plantago, Linn. Plantain.
>major, Linn.
>cordata, Lam.

PRIMULACEÆ. The Primrose Family.

Primula, Linn. Primrose.
>farinosa, Linn. Lake Superior, Dr. Houghton.
>Mistassinica, Michx. Dells of the Wisconsin, Mr. B. F. Mills.

Dodecatheon, Linn.
>Meadia, Linn. Shooting Star.

Trientalis, Linn.
    Americana, Pursh.

Lysimachia, Linn. Loosestrife.
    stricta, Ait. Marquette Co., Mr. John Townley.
    quadiifolia, Linn.
    ciliata, Linn.
    lanceolata, Walt. L. hybrida, Michx. Brookfield, Mr. M. Spears.
    angustifolia, Lam. L. revoluta, Nutt.

Naumburgia, Mœnch.
    thyrsiflora, Reich.

LENTIBULACEÆ. The Bladderwort Family.

Utricularia, Linn.
    purpurea, Walt. Lac Chetac, Dr. Houghton.
    vulgaris, Linn.
    minor, Linn.
    intermedia, Hayne. Brookfield, Mr. M. Spears.
    cornuta, Michx. Lake Superior, Dr. Houghton.

OROBANCHACEÆ. The Broom Rape Family.

Epiphegus, Nutt. Beech Drops.
    Virginiana, Bart.

Conopholis, Wallr. Orobanche, Linn.
    Americana, Wallr.

Aphyllon, Mitchell.
    uniflorum, Tor. &. Gr. Marquettee Co., Mr. J. Townley.

ACANTHACEÆ. The Acanthus Family.

Dipteracanthus, Nees. Ruellia, Linn.
    hybridus, Nees. Beloit, Mr. T. McEl Henry.

SCROPHULARIACEÆ. The Figwort Family.

Verbascum, Linn. Mullen.
    Thapsus, Linn. Introduced.

Linaria, Tourn. Toad Flax.
    vulgaris, Mill. Waukesha, Mr. G. H. Cornwall.

Scrophularia, Linn. Figwort.
    nodosa, Linn. S. Marylandica.

Collinsia, Nutt.
    verna, Nutt.
    parviflora, Dougl.   South shore of Lake Superior, Dr. Z. Pitcher.

Chelone, Tourn.   Snake Head.
    glabra, Linn.

Pentstemon, Mitchell.
    pubescens, Solander.

Mimulus, Linn.   Monkey Flower.
    ringens, Linn.
    Jamesii, Torr. & Gray.

Gratiola, Linn.
    virginiana, Linn.   Wisconsin River, Grant Co.

Synthyris, Benth.
    Houghtoniana, Benth.

Veronica, Linn.   Speedwell.
    Virginica, Linn.   Leptandra, Nutt.
    Angalis, Linn.
    scutellata, Linn.
    peregrina, Linn.

Gerardia, Linn.
    purpurea, Linn.
    tenuifolia, Vahl.
    setacea, Waltr.
    pedicularia, Linn.   Wisconsin River, Prof. Douglass.   Neenah River, Dr.
    grandiflora, Benth.   [Houghton.

Castilleja, Mutis.   Painted Cup.
    coccinea, Spreng.
    sessiliflora, Pursh.   Euchroma grandiflora, Nutt.

Pedicularis, Tourn.   Lousewort.
    Canadensis, Linn.
    lanceolata, Michx.

Melampyrum, Tourn.   Cow Wheat.
    pratense, Linn.   M. Americanum, Michx.

VERBENACEÆ. The Vervain Family.

Verbena, Linn.   Vervain.
    hastata, Linn.
    urticifolia, Linn.
    angustifolia, Linn.
    bracteosa, Michx.
    stricta, Vent.   Upper Mississippi, Dr. Houghton.

Phryma, Linn.   Lopseed.
   leptostachya, Linn.

LABIATÆ.   The Mint Family.

Mentha, Linn.   Mint.
   Canadensis, Linn.   M. borealis, Michx.

Lycopus, Linn.
   Virginicus, Linn.   Bugle Weed.
   sinuatus, Ell.   L. Europæus, Ph.

Micromeria, Benth.
   glabella, Benth.   Hedeoma glabra, Nutt.   Lake Michigan to Lake Supe-
          [rior and Upper Mississippi, Dr. Houghton.

Blephilla, Raf.
   ciliata, Raf.

Monarda, Linn.   Horse Mint.
   fistulosa, Linn.
   punctata, Linn.

Nepeta, Linn.   Catnip.
   cataria, Linn.

Lopanthus, Benth.
   anisatus, Benth.   Falls of St. Croix, Dr. Parry.
   nepetoides, Benth.
   scrophulariæfolius, Benth.

Pycnanthemum, Michx.   Mountain Mint.
   lanceolatum, Pursh.

Prunella, Linn.   Heal-All.
   vulgaris, Linn.

Scutellaria, Linn.   Scullcap.
   parvula, Michx.
   galericulata, Linn.
   lateriflora, Linn.   Mad Dog Scullcap.
   cordifolia, Muhl.

Physostegia, Benth.   Dracocephalum, Linn.
   Virginiana, Benth.   Dragon-Head.

Leonurus, Linn.   Motherwort.
   Cardiaca, Linn.

Galeopsis, Linn.
    tetrahit, Linn.   Lake Superior, Dr. Parry.

Stachys, Linn. Horse Nettle.
    aspera, Michx.

Teucrium, Linn. Germander.
    Canadense, Linn.

BORAGINACEÆ. The Borage Family.

Onosmodium, Michx.
    Virginianum, DC. O. hispidum, Mx.

Lithospermum, Tourn. Gromwell.
    officinale, Linn.
    hirtum, Lehm. Marquette Co., Mr. John Townley.
    canescens, Lehm. Batschia canescens, Mx.

Pentalophus, DC. Batschia, Nutt.
    longiflorus, DC. Beloit, Mr. T. McEl Henry.

Mertensia, Roth. Pulmonaria, Linn. Lungwort.
    Virglnica, DC. Beloit, Mr. McEl Henry.
    Myostis, Linn.
    Stricta, Linn. Marpuette Co., Mr. J. Townley.

Echinospermum, Swartz.
    Lappula, Lehm.

Cynoglossum, Tourn. Hound's-Tongue.
    officinalis, Linn.
    Virginicum, Linn.
    Morisoni, DC.

HYDROPHYLLACEÆ. The Water-Leaf Family.

Hydrophyllum, Linn. Water-Leaf.
    Virginicum, Linn.
    appendiculatum, Michx. Beloit, Mr. T. McEl Henry.

POLEMONIACEÆ.

Polemonium, Tourn.
    reptans, Linn. Jacob's Ladder.

Phlox, Linn.
    glaberrima, Linn. P. revoluta, Eaton.
    pilosa, Linn.
    divaricata, Linn.

CONVOLVULACEÆ.

CALYSTEGIA, R. Brown.   Convolvulus, Linn.
   sepium, R. Brown.
   spithamæa, Pursh.

CUSCUTA, Tourn.
   Gronovii, Willd.   C. Americana, Pursh.

SOLANACEÆ.   THE NIGHTSHADE FAMILY.

DATURA, Linn.   Jamestown Weed.
   Stramonium, Linn.   Introduced.

NICANDRA, Adans.
   physaloides, Gærtn.

PHYSALIS, Linn.   Ground Cherry.
   viscosa, Linn.

SOLANUM, Linn.
   nigrum, Linn.   Night Shade.

GENTIANACEÆ.   THE GENTIAN FAMILY.

GENTIANA, Linn.
   quinqueflora, Lam.
   crinita, Fræl.
   detonsa, Fries.
   rubricaulis, Keating.   Waukesha, Mr. G. H. Cornwall.
   saponaria, Linn.

HALENIA, Borkh.   Swertia, Linn.
   deflexa, Griesb.   Mauvaise and Bois Brule Rivers, Dr. Houghton.

MENYANTHES, Tourn.   Buckbean.
   trifoliata, Linn.

APOCYNACEÆ.   THE DOGBANE FAMILY.

APOCYNUM, Tourn.
   androsæmifolium, Linn.
   hypericifolium, Ait.

ASCLEPIADACEÆ.   THE MILKWEED FAMILY.

ASCLEPIAS, Linn.
   Cornuti, Decaisne.   A. Syriaca, Linn.
   phytolaccoides, Pursh.

(ASCLEPIAS.)

>   variegata, Linn.
>   obtusifolia, Michx.
>   rubra, Linn.   Waukesha, Mr. G. H. Cornwall.
>   incarnata, Linn.
>   verticillata, Linn.   Beloit, Dr. S. P. Lathrop.
>   tuberosa, Linn.   Butterfly Weed.
>   lanuginosa, Nutt.   Eagle Prairie, I. A. Lapham.

ACERATES, Ell.

>   viridiflora, Ell.
>   longifolia, Ell.   Marquette Co., Mr. J. Townley.

OLEACEÆ.   THE OLIVE FAMILY.

FRAXINUS, Tourn.   Ash.

>   Americana, Linn.   White Ash.
>   sambucifolia, Lam.   Black Ash.

ARISTOLOCHIACEÆ.   THE BIRTHWORT FAMILY.

ASARUM, Tourn.

>   Canadense, Linn.   Colt's Foot.

CHENOPODIACEÆ.   THE GOOSEFOOT FAMILY.

SALSOLA, Linn.   Saltwort.

>   kali, Linn.   Lake Shore, Milwaukee.

CHENOPODIUM, Linn.

>   album, Linn.   Pigweed.
>   hybridum, Linn   Goosefoot.

BLITUM, Tourn.

>   capitatum, Linn.   Indian Strawberry.

AMBRINA, Spach.

>   Botrys, Moquin.   Brookfield, Mr. G. H. Cornwall.

ACNIDA, Mitchell.   Water Hemp.

>   cannabina, Linn.   St. Croix, Dr. Parry.

AMARANTHACEÆ.   THE AMARANTH FAMILY.

AMARANTHUS, Linn.

>   altissimus, Riddell.
>   hybridus, Linn.   Red Root.
>   hypocondriacus, Linn.   Prince's Feather.
>   tamariscinus, Nutt.

NYCTAGINACEÆ. The Nyctago Family.

Allionia, Linn.

    albida, Wallr. St Croix River, Dr. Houghton.
    nyctaginea, Michx. St. Croix River, Dr. Houghton.

Oxybaphus angustifolius, Torr. St. Croix, Dr. Parry.

POLYGONACEÆ. The Buckwheat Family.

Polygonum, Linn.

    Pennsylvanicum, Linn.
    Persicaria, Linn.
    Hydropiper, Linn. P. punctatum, Ell.
    hydropiperoides, Michx.
    amphibium, Linn.
    aviculare, Linn. Knot-Grass.
    articulatum, Linn. Lakes Michigan and Superior, Dr. Houghton.
    Virginianum, Linn.
    sagittatum, Linn.
    Convolvulus, Linn.
    cilinode, Michx. Lake Superior, Dr. Houghton.
    dumetorium, Linn. P. scandens, L.

Rumex, Linn. Dock.

    Britannica, Linn.
    crispus, Linn. Curled Dock.
    Acetosella, Linn. Sorrel.
    hydrolapathum, Hudson. St. Croix. Dr. Parry.

THYMELEACEÆ. The Mezereum Family.

Dirca, Linn. Leatherwood.

    palustris, Linn.

ELÆAGNACEÆ. The Oleaster Family.

Shepherdia, Nutt.

    Canadensis, Nutt.

SANTALACEÆ. The Sandal Wood Family.

Comandra, Nutt.

    umbellata, Nutt.

ULMACEÆ. The Elm Family.

Ulmus, Linn. Elm.

    Americana, Linn. White Elm.
    fulva, Michx. Slippery Elm.

CELTIS, Tourn.
    occidentalis, Linn. Hack Berry.

SAURURACEÆ. THE LIZARD'S TAIL FAMILY.

SAURURUS, Linn. Lizard's Tail.
    cernuus, Linn. Upper Mississippi, Dr. Houghton.

CALLITRICHACEÆ.

CALLITRICHE, Linn.
    verna, Linn.

EUPHORBIACEÆ. THE SPURGE FAMILY.

EUPHORBIA, Linn.
    corollata, Linn.
    maculata, Linn.
    hypericifolia, Linn. St. Croix, Dr. Parry.

EMPETRACEÆ. THE CROWBERRY FAMILY.

EMPETRUM, Tourn. Crowberry.
    nigrum, Linn. Lake Superior, Dr. Houghton.

JUGLANDACEÆ. THE WALNUT FAMILY.

JUGLANS, Linn. Walnut.
    cinerea, Linn. Butternut or White-Walnut.
    nigra, Linn. Black-Walnut.

CARYA, Nutt. Hickory.
    alba, Nutt. Shag-Bark Hickory.
    glabra, Torr. C. porcina, Nutt. Pignut Hickory.

CUPULIFERÆ. THE OAK FAMILY.

QUERCUS, Linn. Oak.
    alba, Linn. White Oak.
    obtusilpba, Michx. Q. stellata, Willd. Post Oak. Upper Mississippi,
    macrocarpa, Michx. Burr Oak.   [Dr. Houghton.
    bicolor, Willd. Swamp White Oak.
    Prinos, Linn. Swamp Chestnut Oak. Near Janesville, I. A. L.
    rubra, Linn. Red Oak.
    palustris, Linn. Pin Oak.

FAGUS, Tourn. Beech.
    ferruginea, Ait.

Corylus, Tourn. Hazel-Nut.
    Americana, Walt.
    rostrata, Ait.

Carpinus, Linn. Hornbeam.
    Americana, Michx.

Ostrya, Micheli. Iron Wood.
    Virginica, Willd.

MYRICACEÆ. The Sweet Gale Family.

Myrica, Linn. Sweet Gale.
    Gale, Linn. Lake Superior, Dr. Houghton.

Comtonia, Solander. Sweet Fern.
    asplenifolia, Ait. Dells of Wisconsin River.

BETULACEÆ. The Birch Family.

Betula, Tourn. Birch.
    papyracea, Ait. Canoe Birch.
    pumila, Linn. B. glandulosa, Mx. Low Birch.

Alnus, Tourn. Alder.
    incana, Willd. A. glauca, Mx. St. Croix River, Dr. Houghton.
    serrulata, Willd.

SALICACEÆ. The Willow Family.

Salix, Tourn. Willow.
    candida, Willd.
    tristis, Ait.
    humilis, Marshall. S. Muhlenbergiana, Barr.
    discolor, Muhl.
    eriocephala, Michx. S. prinoides, Ph. Mauvaise River, Dr. Houghton.
    sericea, Marshall. S. Grisea, Willd.
    rostrata, Richardson.
    lucida, Muhl.
    longifolia, Muhl. Upper Mississippi, Dr. Houghton.
    pedicellaris, Pursh. St. Croix, Dr. Parry.

Populus, Tourn. Poplar.
    tremuloides, Michx. Quaking Aspen.
    grandidentata, Michx.
    candicans, Ait. Balm of Gilead.

PLATANACEÆ. The Plane-Tree Family.

Platanus, Linn. Sycamore, Buttonwood.
    occidentalis, Linn.

URTICACEÆ. THE NETTLE FAMILY.

HUMULUS, Linn. Hop.
    Lupulus, Linn.

URTICA, Tourn. Nettle.
    dioica, Linn.
    Canadensis, Linn.

PILEA, Lindl. Adike, Raf.
    pumila, Linn.

PARIETARIA, Tourn. Pellitory.
    Pennsylvanica, Muhl.

CONIFERÆ. THE PINE FAMILY.

PINUS, Tourn. Pine.
    Banksiana, Lambert. Dells of the Wisconsin.
    resinosa, Ait. Red Pine.
    mitis, Michx. Yellow Pine. Dane County, I. A. L.
    Strobus, Linn. White Pine.

ABIES, Tourn. Spruce.
    balsamea, Marsh. Balsam Fir. Manitowoc, I. A. L.
    Canadensis, Michx. Hemlock. Manitowoc, I. A. L.
    nigra, Poir. Black Spruce. Lake Superior, Dr. Houghton. Upper
                                 [Mississippi, Prof. Douglass.
    alba, Michx. White Spruce. Upper St. Croix, Dr. Parry.

LARIX, Tourn. Larch.
    Americana, Michx. Tamarack.

THUJA, Tourn.
    occidentalis, Linn. White Cedar.

JUNIPERUS, Linn. Juniper.
    communis, Linn.
    Virginiana, Linn. Red Cedar.

TAXUS, Tourn. Yew.
    Canadensts, Willd.

ARACEÆ. THE ARUM FAMILY.

ARUM, Linn. Indian Turnip.
    triphyllum, Linn.

CALLA, Linn. Water Arum.
    palustris, Linn.

Symplocarpus, Salisb.   Skunk Cabbage.
   foetidus, Salisb.

Acorus, Linn.   Sweet Flag.
   calamus, Linn.

LEMNACEÆ.   The Duckweed Family.

Lemna, Linn.   Duckweed.
   minor, Linn.
   trisulca, Linn.

TYPHACEÆ.   The Cat-Tail Family.

Typha, Tourn.   Cat-Tail.
   latifolia, Linn.

Sparganium, Tourn.
   ramosum, Hudson.
   natans, Linn.   St. Croix, Dr. Parry.

NAIDACEÆ.   The Pondweed Family.

Potamogeton, Tourn.   Pondweed.
   amplifolius, Tuckerman.
   perfoliatus, Linn.
   compressus, Linn.   P. zosterifolius, Schum.
   pauciflorus, Pursh.   P. gramineus, Mx.
   pectinatus, Linn.
   heterophyllns, Schreber.

ALISMACEÆ.   The Water Plantain Family.

§ 1. *Juncagineæ.*

Triglochin, Linn.
   elatum, Nutt.

Scheuchzeria, Linn.
   palustris, Linn.

§ 2. *Alismeæ.*

Alisma, Linn.   Water Plantain.
   Plantago, Linn.

Echinodorus, Richard.
   subulatus, Engleman.   St. Croix, Dr. Parry.

Sagittaria, Linn.
   sagittifolia, Linn.   Arrow-Leaf.

HYDROCARDIACEÆ. The Frog's-Bit Family.

Udora, Nutt. Water Weed.
>Canadensis, Nutt.

Vallisneria, Mitcheli. Tape Grass.
>spiralis, Linn.

ORCHIDACEÆ. The Orchis Family.

Microstylis, Nutt.
>monophyllas, Lindl. St. Croix, Dr. Parry.
>ophioglossoides, Nutt.

Liparis, Richards.
>liliifolia, Richards.
>Lœselii, Richards. Waukesha, Mr. G. H. Cornwall.

Corallorhiza, Haller.
>multiflora, Nutt. Waukesha. Mr. G. H. Cornwall.

Aplectrum, Nutt. Putty Root.
>hyemale, Nutt.

Orchis, Linn.
>spectabilis, Linn.

Platanthera, Richards.
>orbiculata, Lindl.
>Hookeri, Lindl.
>bracteata, Torr.
>dilatata, Lindl. St. Croix, Dr. Parry.
>hyperborea, Lindl.
>leucophæa, Nutt.
>psycodes, Gray. O. fimbriata Pursh.

Arethusa, Gronov.
>bulbosa, Linn. Lemonwier River.

Pogonia, Juss.
>ophioglossoides, Nutt. Waukesha, Mr. G. H. Cornwall.

Calopogon, R. Brown.
>pulchellus, R. Br.

Spiranthes, Richards. Lady's Tresses.
>gracilis, Bigelow.
>cernua, Richards

Goodyera. R. Brown.
>pubescens, R. Brown.

Cypripedium, Linn. Lady's Slipper.
  pubescens, Willd. Yellow Lady's Slipper.
  parviflorum, Salisb. Small Yellow Lady's Slipper.
  candidum, Muhl. White Lady's Slipper.
  spectabile, Swartz. Moccason Flower.
  acaule, Ait. Purple Lady's Slipper.

AMARYLLIDACEÆ. The Amaryllis Family.

Hypoxis, Linn. Star Grass.
  erecta, Linn.

HÆMODORACEÆ. The Bloodwort Family.

Aletris, Linn. Star Grass.
  farinosa, Linn. Marquette Co., Mr. J. Townley.

IRIDACEÆ. The Iris Family.

Iris, Linn.
  versicolor, Linn. Blue Flag.
  lacustris, Nutt.

Sisyrinchium, Linn. Blue-eyed Grass.
  Bermudianum, Linn.

DIOSCORACEÆ. The Yam Family.

Dioscorea, Plumier.
  villosa, Linn.

SMILACEÆ. The Green Brier Family.

§ 1. *Smilaceæ.*

Smilax, Tourn.
  rotundifolia, Linn. Green Brier.
  herbacea, Linn.
  lasioneuron, Hook.

§ 2. *Trilliaceæ.*

Trillium, Linn.
  cernuum, Linn. T. pendulum, Muhl.
  grandiflorum, Salisb.
  nivale, Riddell.
  recurvatum, Beck.
  sessile, Linn.

Medeola, Gronov.
  Virginica, Linn.

LILIACEÆ. The Lily Family.

§ 1. *Asparageæ.*

Asparagus, Linn.
officinale, Linn.   Introduced.

Polygonatum, Tourn.   Solomon's Seal.
pubescens, Pursh.

Smilacina, Desf.
racemosa, Desf.
stellata, Desf.
trifolia, Desf.
bifolia, Ker.

Clintonia, Raf.
borealis, Raf.

§ 2. *Asphodeleæ.*

Scilla, Linn.
esculenta, Ker.   Beloit, Dr. S. P. Lathrop.

Allium, Linn.   Garlic.
Canadense, Kalm.
cernuum, Roth.
tricoccum, Ait.   Leek.

§ 3. *Tulipaceæ.*

Lilium, Linn.   Lily.
Philadelphicum, Linn.   Orange Lily.
Canadense, Linn.   Nodding Lily.

Erythronium, Linn.   Dog's Tooth Violet.
Americanum, Smith.
albidum, Nutt.

MELANTHACEÆ. The Colchicum Family.

§ 1. *Uvularieæ.*

Uvularia, Linn.   Bellwort.
grandiflora, Smith.
sessilifolia, Linn.   Marquette Co., Mr. John Townley.

Streptopus, Michx.
roseus, Michx.

§ 2. *Melanthieæ.*

Zygadenus, Michx.
glaucus, Nutt.

Tofieldia, Hudson.
glutinosa, Willd.

JUNCACEÆ. The Rush Family.

Luzula, DC. Woodrush.
pilosa, Willd.
campestris, Linn.

Juncus, Linn. Rush.
effusus, Linn.
Balticus, Willd.
scirpoides, Lam. T. polycephalus, Mx.
acuminatus, Michx.
tenuis, Willd.
conradi, Tuckerman. St. Croix, Dr. Parry.

PONTEDERIACEÆ. The Pickerel-Weed Family.

Pontederia, Linn. Pickerel-Weed.
cordata, Linn.

COMMELYNACEÆ. The Spiderwort Family.

Tradescantia, Linn. Spiderwort.
Virginica, Linn.

CYPERACEÆ. The Sedge Family.

Cyperus, Linn.
diandrus, Torr.
inflexus, Muhl. Bass Lake, Dane Co., I. A. Lapham.
schweinitzii, Torr. St. Croix, Dr. Parry.
strigosus, Linn.
filiculmis, Vahl. Upper Mississippi, Dr. Houghton.

Dulichium, Richard.
spathaceum, Pers.

Eleocharis, R. Brown.
obtusa, Schultz.
palustris, R. Brown.
tenuis, Schultz.
acicularis, R. Brown.

Scirpus, Linn.
pungens, Vahl. S. triqueter, Mx.
lacustris, Linn. Bulrush.
fluviatilis, Gray. S. maratimus, var. Torr.
atrovirens, Muhl.
lineatus, Michx.
Eriophorum, Michx.

Eriophorum, Linn.   Cotton Grass.

    alpinum, Linn.   Lake Superior, Dr. Houghton.
    vaginatum, Linn.
    Virginicum, Linn.
    polystachyum, Linn.
    angustifolium, Richard.

Scleria, Linn.

    triglomerata, Mx.   Beloit, Dr. S. P. Lathrop.

Carex, Linn,   Ledge.

    aurea, Nutt.
    anceps, Willd.
    blanda, Dewey.
    bromoides, Schk.
    Buxbaumii, Wahl.
    bullata, Schk.
    chordorrhiza, Ehrh.
    comosa, Boott.
    Deweyana, Schw.
    eburnea, Boott.
    festucacea, Schk.
    gracilis, Ehrh.   C. disperma, Dewey.
    granularis, Muhl.
    gracillima, Schw.
    grisea, Wahl.   Beloit, Dr. S. P. Lathrop.
    hystriciana, Willd.
    intermedia, Good.
    irrigua, Smith.
    intumescens, Rudge.
    laxiflora, Lam.
    lanuginosa. Michx.   C. Pellita, Muhl.
    lacustris, Willd.
    lupulina, Muhl.
    longirostris, Torr.   Blue Mounds, I. A. Lapham.
    miliacea, Muhl.   Beloit, Dr. S. P. Lathrop.
    mirabilis, Dew.
    oligosperum, Michx.
    plantaginea, Lam.
    polytrichoides, Muhl.
    panicea, Linn.   C. Meadii, Dew.
    Pennsylvanica, Lam.
    pubescens, Muhl.   Brookfield, Mr. M. Spears.
    rosea, Schk.
    rigida, Good.   C. saxatalis.   Lake Superior, Dr. Houghton.
    siccata, Dewey.   Beloit, Dr. S. P. Lathrop.
    stipata, Muhl.
    sparganoides, Nuhl.
    stellulata, Good.

(Carex.)

straminea, Schk.
stricta, Lam.   C. acuta, Muhl.   C. angustata, Boott.
teretiuscula, Good.
tenera, Dew.
vulpinoidea, Michx.   C. multiflora, Muhl.

GRAMINEÆ.   The Grass Family.

Leersia, Solander.   White Grass.

oryzoides, Swartz.   Cut Grass.
Virginica, Willd.   White Grass.

Zizania, Gronov.   Wild Rice.

aquatica, Linn.

Alopecurus, Linn.   Fox-Tail Grass.

aristulatus, Mx.

Phleum, Linn.   Timothy.

pratense Linn.

Agrostis, Linn.   Bent-Grass.

scabra, Willd.   Thin Grass.   A. Michauxii, Trin.
vulgaris, With.   Redtop.

Cinna, Linn.

arundinacea, Linn.

Muhlenbergia, Schreber.

glomerata, Trin.   Polypogon racemosa, Nutt.
Mexicana, Trin.   Agrostis lateriflora, Mx.
Willdenovii, Trin.   Agrostis tenuiflora, Willd.

Brachyelytrum, Beauv.

aristatum, Beauv.   Muhlenbergia erecta, Schreb.

Calamagrostis, Adans.

Canadensis, Beauv.

Oryzopsis, Michx.   Mountain Rice.

asperifolia, Michx.
melanocarpa, Muhl.   Piptatherum nigrum, Torr.

Stipa, Linn.

avenacea, Linn.

Aristida, Linn.

tuberculosa, Nutt.   St. Croix, Dr. Parry.

SPARTINA, Schreb. Cord Grass.
cynosuroides, Willd.

BOUTELOUA, Lagasca.   Athcropogon, Muhl.
racemosa, Lag.   A. apludoides, Michx.
papillosa, Gray.   Cassville, Dr. Houghton.

KŒLERIA, Pers.
cristata, Pers.   K. nitida, Nutt.

REBOULEA, Kunth.
obtusata, Kœleria truncata Torr.

MELICA, Linn.   Melic-Grass.
speciosa, Muhl.   Beloit, Dr. S. P. Lathrop.

GLYCERIA, R. Brown.
Canadensis, Trin.   Poa Canadensis.
nervata, Trin. Poa nervata, Willd.
fluitans, R. Brown.
aquatica, Smith.

POA, Linn.   Meadow Grass.
compressa, Linn. Blue Grass. Beloit, Dr. S. P. Lathrop. Introduced.
debilis, Torr.
nemoralis, Linn.
serotina, Ehrh.
trivialis, Linn.
pratensis, Linn.

ERAGROSTIS, Beauv.
megastachya, Link.   Poa eragrostis, Linn.

FESTUCA, Linn.
nutans, Willd.
ovina, Linn.   Two Rivers, Manitowoc Co.

BROMUS, Linn.   Brome Grass.
ciliatus, Linn.
purgans, Linn.
secalinus, Linn.   Chess.

PHRAGMITES, Trin.   Reed.
communis, Trin.

TRITICUM, Linn.   Wheat.
repens, Linn.   Couch-Grass.
dasystachyum, Gray.   Two Rivers.
caniuum, Linn.   Beloit, Dr. S. P. Lathrop.

Elymus, Linn. Lime-Grass.
   Virginicus, Linn.
   Canadensis, Linn.
   glaucifolius, Muhl.
   striatus, Willd.
   hystrix, Linn. Bottle-Brush Grass.

Hordeum, Linn. Barley.
   jubatum, Linn. Squirrel-Tail Grass.

Aira, Linn. Hair-Grass.
   cæspitosa, Linn.

Danthonia, DC. Wild-Oat Grass.
   spicata, Beauv.

Avena, Linn. Oat.
   striata, Michx. Trisetum purpurascens, Torr.

Holcus, Linn. Velvet Grass.
   lanatus, Linn. Upper Mississippi, Dr. Houghton.

Hierochloa, Gmelin. Seneca Grass.
   borealis, Rœm. & Sch.

Milium, Linn. Millet-Grass.
   effusum, Linn.

Panicum, Linn. Panic Grass.
   capillare, Linn.
   virgatum, Linn.
   latifolium, Linn.
   clandestinum, Linn.
   dichotomum, Linn.
   pubescens, Lam.
   crus-galli, Linn. Barn-Yard Grass.
   longisetum, Torr. Neenah River, Prof. Douglass.
   xanthophysum, Gray. Beloit. Dr. S. P. Lathrop.
   depauperatum, Muhl. Beloit, Dr. S. P. Lathrop.

Setaria, Beauv.
   glauca, Beauv.

Cenchrus, Linn. Burr-Grass.
   tribuloides, Linn. Upper Mississippi, Dr. Houghton.

Andropogon, Linn. Beard-Grass.
   furcatus, Muhl.
   scoparius, Michx.

Sᴏʀɢʜᴜᴍ, Pers.

    nutans. Andropogon nutans, Linn.

EQUISETACEÆ. Tʜᴇ Hᴏʀsᴇ-Tᴀɪʟ Fᴀᴍɪʟʏ.

Eǫᴜɪsᴇᴛᴜᴍ, Linn. Horse-Tail.
    arvense, Linn.
    eburnum, Schreb. Lake Superior, Dr. Torrey.
    sylvaticum, Linn.
    limosum, Linn.
    hyemale, Linn. Scouring Rush.
    lævigatum, Braun.
    variegatum, Schleicher.
    scirpoides, Michx.

FILICES. Tʜᴇ Fᴇʀɴ Fᴀᴍɪʟʏ.

Pᴏʟʏᴘᴏᴅɪᴜᴍ, Linn.

    vulgare, Linn. Blue Mounds, I. A. Lapham.
    phegopteris, Linn. St. Croix, Dr. Parry.
    dryopteris, Linn. Dells of the Wisconsin.

Sᴛʀᴜᴛʜɪᴏᴘᴛᴇʀɪs, Willd. Ostrich Fern.

    Germanica, Willd.

Aʟʟᴏsᴏʀᴜs, Bernhardi.

    gracilis, Presl. Dells of the Wisconsin.

Asᴘɪᴅɪᴜᴍ.

    fragrens, Sw. Falls of the St. Croix, Dr. Parry.

Aᴅɪᴀɴᴛᴜᴍ, Linn. Maiden Hair.

    pedatum, Linn.

Pᴛᴇʀɪs, Linn. Brake.
    aquilina, Linn.
    atropurpurea, Linn.

Cʜᴇɪʟᴀɴᴛʜᴇs, Swartz. Lip Fern.

    vestita, Willd. Falls of St. Croix, Dr. Parry.

Cᴀᴍᴘᴛᴏsᴏʀᴜs, Link. Asplenum, Linn.

    rhizophyllus, Link. Walking Leaf.

Asᴘʟᴇɴɪᴜᴍ, Linn. Spleenwort.

    trichomanes, Linn. Dells of the Wisconsin.
    thelypteroides, Michx.
    Filix-femina, R. Brown. Aspidium asplenoides,

Cʏsᴛᴏᴘᴛᴇʀɪs, Bernhardi. Bladder Fern.
    bulbifera, Bernh.

Woodsia, R. Brown.
   ilvensis, R. Brown.   Blue Mounds, I. A. Lapham.

Dryopteris, Adans.   Aspidium, Linn.   Wood Fern.
   thelypteris, Gray.
   dilatata, Gray.   Falls of the St. Croix, Dr. Parry.
   cristata, Gray.
   Goldiana, Gray.

Onoclea, Linn.   Sensitive Fern.
   sensibilis, Linn.

Osmunda, Linn.   Flowering Fern.
   spectabilis, Willd.
   Claytoniana, Linn.   O. interrupta, Mx.
   cinnamomea,   Linn.

Botrychium, Swartz.
   lunarioides, Sw.   B. fumarioides, Willd.
   Virginicum, Sw.

LYCOPODIACEÆ.   The Club Moss Family.

Lycopodium, Linn.   Club Moss.
   lucidulum, Michx.
   annotinum, Linn.   Lake Superior, Dr. Houghton.
   dendroideum, Michx.   Lake Superior to Upper Mississippi, Dr.
   clavatum, Linn.   [Houghton.
   complanatum, Linn.

Selaginella, Beauv.
   rupestris, Spring.   Lyc. rupestre, Linn.   Blue Mounds.
   apus, Spring.   L. apodum, Linn.

CHARACEÆ.   The Chara Family.

Chara, Linn.
   vulgaris, Willd.   Feather-Beds.

MUSCI.   Mosses.

Funaria, Schreber.
   hygrometrica, Hedwig.   Lake Superior, Dr. Parry.

Dicranum, Hedwig.
   scoparium, Hedwig.

Leucobryum, Hampe.
   vulgare, Hampe.   St. Croix, Dr. Parry.

ATRICHUM, Beauv.
    angustatum, Beauv.   Lake Superior, Dr. Parry.

BARTRAMIA, Hedw.
    pomiformis, Hedw.   Montreal River, Dr. Parry.

MINUM, ——
    affine, Blandon.   Lake Superior, Dr. Parry.

BRYUM, Linn.
    roseum, Schreber.   Lake Superior, Dr. Parry.

HYPNUM, Linn.   Feather Moss.
    populum, Hedw.   Lake Superior, Dr. Parry.
    Schreberi, Willd.   Lake Superior, Dr. Parry.
    tamariscinum, Hedw.   Lake Superior, Dr. Parry.

CLIMACIUM, Weber & Mohr.   Tree Moss.
    dendroides, W. & M.   Lake Superior, Dr. Parry.

HEPATICÆ.   LIVERWORTS.

RICCIA, Mitchell.
    natans, Linn.

MARCHANTIA, Linn.
    polymorpha, Linn.

LICHENES.

CLADONIA, Hoffm.   Reindeer Moss.
    rangiferina, Hoffm.   Falls of St. Croix, Dr. Parry.

UMBILICARIA, Hoffm.   Tripe de Roche.
    Muhlenbergii, Ach.   Falls of St. Croix, Dr. Parry.

WOODS OF WISCONSIN.

RACINE, November 25, 1852.

DEAR SIR:—Your letter, inviting me to prepare a paper for the State Agricultural Society, upon the "Woods of Wisconsin," and making suggestions as to the manner in which the subject should be considered, came duly to hand.

In reply, I cannot but say, that were I to consult my other engagements, I should find it impossible to comply with your request; yet in view of the importance of the subject you suggest, and believing also, as I do, that the State has a claim upon every citizen to contribute his part, so far as able, to the fund of information on any subjects immediately important to that class whose prosperity is of vital importance to the welfare of all. In view of these considerations, I will endeavor, in a brief manner, to comply with your request, so far as possible, in the short time given.

It will not be my purpose, in this communication, to give a botanical description, but merely brief notes on the principal qualities and value of the woods, and the fitness of the various trees for the purposes of ornament. I take great pleasure in tendering my thanks, to a much esteemed friend, and accomplished botanist, I. A. Lapham, Esq., of Milwaukee, for his Catalogue of the Flora of Wisconsin,* which embraces a more complete list of our Forest Trees, than can elsewhere be found.

TREES INDIGENOUS TO WISCONSIN.

OAKS.

This family bears transplanting rather poorly, unless quite small. They are readily raised from seed, which should be kept in a cool, dry place, till March or April, when they must be planted, two inches deep, in rich vegetable mould.

WHITE OAK—*Quercus Alba*.

This noble tree is the largest and most important of the American oaks. The excellent properties of the wood render it eminently valuable for a great variety of uses. Wherever strength and durability is required, the White Oak stands in the first rank. It is employed in making wagons, coaches and sleds; staves and hoops of the best quality for barrels and casks, are obtained from this tree; it is extensively used in architecture, ship-building, &c., and vast quantities are used for fencing. The bark is employed in tanning. The domestic consumption of this tree is so great that it is of the first importance to preserve the young trees wherever it is practicable, and to make young plantations where the tree is not found. The White Oak is a graceful ornamental tree, and worthy of particular attention as such—found abundantly in most of the timbered districts.

---

*See ante, page 337.

Burr Oak—*Q. Macrocarpa.*

This is perhaps the most ornamental of our oaks. Nothing can exceed the graceful beauty of these trees, when not crowded or cramped in their growth, but left free to follow the laws of their development. Who has not admired these trees in our extensive Burr Oak openings? Its large leaves are a dark-green above, and a bright silvery-white beneath, which gives the tree a singularly fine appearance when agitated by the wind. The wood is tough, close-grained, and more durable than the White Oak, especially when exposed to frequent changes of moisture and drying; did the tree grow to the same size, it would be preferred for most uses. Abundant, and richly worthy of cultivation both for utility and ornament.

Swamp White Oak—*Q. Bicolor.*

Another valulable and ornamental oak, rather larger than the Burr Oak. The wood is close-grained, durable, splits freely, and is well worthy of cultivation. Not quite so common as the Burr Oak. Valuable for fuel.

Post Oak—*Q. Obtusiloba.*

A scraggy, small tree, found sparingly in this State. The timber is durable, and makes good fuel. Not worthy of cultivation.

Swamp Chestnut Oak—*Q. Prinos.*

This species of Chestnut Oak is a large, graceful tree, wood rather open-grained, yet valuable for most purposes to which the oaks are applied; makes the best fuel of any of this family. A rare tree, found by Mr. Lapham, growing near Janesville. Worthy of cultivation.

Red Oak—*Q. Rubra.*

The Red Oak is a well known, common and rather large tree. The wood is coarse-grained, and the least durable of the oaks, nearly worthless for fuel, and scarcely worthy of cultivation, even for ornament.

Pin Oak—*Q. Palustris.*

One of the most common trees, in many sections of the State. The wood is of little value, even for fuel. The tree is quite ornamental, and should be sparingly cultivated for this purpose.

MAPLES.

A family of beautiful trees, which bear transplanting well; even those of from six to eight inches in diameter, can, without difficulty, be successfully transplanted. Grows in almost any well-drained soil, free from stagnant water. The seed should be planted in the fall, as soon as ripe; they require but a slight covering.

Sugar Maple—*Acer Saccharinum*.

This well-known and noble tree is found growing abundantly in many sections of the State. The wood is close-grained and susceptible of a beautiful polish, which renders it valuable for many kinds of furniture, more especially the varieties known as Bird's-eye and Curled Maples.— The wood lacks the durability of the oak; consequently, is not valuable for purposes where it will be exposed to the weather. For fuel, it ranks next to Hickory. The sugar manufactured from this tree affords no inconsiderable resource for the comfort and even wealth of many sections of the northern States, especially those newly settled, where it would be difficult and expensive to procure their supply from a distance. As an ornamental tree, it stands almost at the head of the catalogue. The foliage is beautiful, compact, and free from the attacks of insects. It puts forth its yellow blossoms early, and in the autumn the leaves change in color, and show the most beautiful tints of red and yellow long before they fall.— Worthy of especial attention for fuel and ornament—well adapted for street planting.

Red Maple—*A. Rubrum*.

Another fine maple, of more rapid growth than the foregoing species. With wood rather lighter, but quite as valuable for cabinet-work—for fuel not quite so valuable. The young trees bear transplanting even better than other maples. Though highly ornamental, this tree hardly equals the first named species. It puts forth, in early spring, its scarlet blossoms before a leaf has yet appeared. Well adapted for street planting.

Mountain Maple—*A. Spicatum*.

A small, branching tree, or rather shrub, found growing in clumps.— Not worthy of much attention.

Box Maple—*Negundo Aceroides.*

This tree is frequently called Box Elder. It is of a rapid growth; quite ornamental. The wood is not used in the arts, but is good fuel.— Should be cultivated; grows on Sugar and Rock rivers.

ELMS.

These are tall, fine trees, more remarkable for ornament than for the value of their wood. They grow to the greatest perfection in deep, moist soils, but will flourish to a considerable extent in almost any productive grounds. They endure transplanting admirably. The seed should be sown immediately after ripening, which is in June.

White Elm—*Ulmus Americana.*

This large and graceful tree stands confessedly at the head of the list of ornamental deciduous trees. Its wide-spreading branches and long pendulous branchlets form a beautiful and conspicuous head. It grows rapidly; is free from disease and the destructive attacks of insects; will thrive on most soils; and for planting along streets, in public grounds or lawns, is unsurpassed by any American tree. The wood is but little used in the arts; makes good firewood; should be planted along all the roads and streets, near every dwelling and on all public grounds.

Slippery Elm—*U. Fulva.*

This smaller and less ornamental species is also common. The wood, however, is much more valuable than the White Elm, being durable and splitting readily. It makes excellent rails, and is much used for the frame work of buildings. Valuable for fuel; should be cultivated.

CHERRY.

A valuable class of trees, which flourish best in a deep, moist, sandy soil. They all bear transplanting well. The seed should be planted immediately when ripe, which is the case with all stony seeds.

Wild Black Cherry—*Cerasus Serotina.*

This large and beautiful species of Cherry is one of the most valuable of American trees. The wood is compact, fine-grained, and of a brilliant, reddish color, not liable to warp, or shrink and swell with atmospheric changes; extensively employed by cabinet makers for every species of furniture; is next in value to mahogany. It is exceedingly durable; hence is valuable for fencing, building, &c. Richly deserving of a place in the lawn or timber plantation.

BIBD CHERRY—*C. Pennsylvanica.*

A small northern species, common in the State, but scarcely worthy of cultivation, unless for ornament.

CHOKE CHERRY—*C. Virginiana.*

This diminutive tree is of little value, not worth the trouble of cultivation.

WILD PLUM—*Prunus Americana.*

The common Wild Plum, when in full bloom, is one of the most ornamental of small flowering trees, and as such, should not be neglected. The fruit is rather agreeable, but not to be compared to fine cultivated varieties, which may be engrafted on the wild stock to the very best advantage. It is best to select small trees, and work them on the roots. The grafts should be inserted about the middle of April.

HACK BERRY—*Celtis Occidentalis.*

An ornamental tree of medium size; wood hard, close-grained, and elastic; makes the best of hoops, whip-stalks, and thills for carriages. The Indians formerly made great use of the Hackberry wood for their bows. A tree worthy of a limited share of attention.

AMERICAN LINDEN, OR BASS WOOD—*Tilia Americana.*

One of the finest ornamental trees for public grounds, parks, &c., but will not thrive where the roots are exposed to bruises; for this reason, it is not adapted to planting along the streets of populous towns. The wood is light and tough; susceptible of being bent to almost any curve; durable if kept from the weather; takes paint well and is considerably used in the medicinal arts; for fuel, nearly worthless. This tree will flourish in almost any moderately rich, damp soil; bears transplanting well; can be propagated readily from layers. The seed should be sown in autumn.

WHITE THORN—*Cratægus Coccinea.*
DOTTED THORN—*C.Punctata.*

These two species of Thorn are found everywhere on the rich bottom lands. When in bloom they are beautiful, and should be cultivated for ornament. The wood is remarkably compact and hard, and were it not for the small size of the tree, would be valuable.

CRAB APPLE—*Pyrus Coronaria*.

This common small tree is attractive when covered with its highly fragrant rose-colored blossoms. Wood hard, fine compact grain, but the tree is too small for the wood to be of much practical value. Well worthy of a place in extensive grounds.

MOUNTAIN ASH—*P. Americana*.

This popular ornament to our yards, is found growing in the northern part of the State, and as far south as 43° lat. The wood is useless.

ASH—*Fraxinus*.

This genus comprises several tall, straight trees, with valuable wood and handsome foliage. It mostly grows in rich, damp soil. It bears transplanting well. The seed may be planted either in the fall or spring.

WHITE ASH—*Fraxinus Acuminata*.

A large, interesting tree, which combines utility with beauty in an eminent degree. The wood possesses strength, suppleness and elasticity, which renders it valuable for a great variety of uses. It is extensively employed in carriage manufacturing; for various agricultural implements; is esteemed superior to any other wood for oars; excellent for fuel. The White Ash grows rapidly, and in open ground forms one of the most lovely trees that is to be found. The foliage is clean and handsome, and in the autumn turns from its bright green to a violet purple hue, which adds materially to the beauty of our autumnal sylvan scenery. It is richly deserving our special care and protection, and will amply repay all labor and expense bestowed on its cultivation.

BLACK ASH—*F. Sambucifolia*.

Another tall, graceful and well known species of Ash. The wood is used for making baskets, hoops, &c.; when thoroughly dry, affords a good article of fuel. Deserves to be cultivated in low, rich, swampy situations, where more useful trees will not thrive.

WALNUT.

The Walnut Family includes the Hickories with the Walnuts proper. The latter bear transplanting well, while the former require great care.

A rich, deep, sandy loam is required to develope the trees in all their luxuriant grandeur. The nuts may either be planted in the autumn or kept in a dry place until spring, when they should be planted three inches deep, in a light vegetable mould.

Black Walnut—*Juglans Nigra.*

This giant of the rich alluvial bottom lands, claims special attention for its valuable timber. It is among the most durable and beautiful of American woods; susceptible of a fine polish; not liable to shrink and swell by heat and moisture. It is extensively employed by the cabinet-makers, for every variety of furniture. Walnut forks are frequently found, which for richness and beauty, rival the far-famed Mahogany. This tree, in favorable situations, grows rapidly; is highly ornamental, and produces annually an abundant crop of nuts.

Butternut—*J. Cinerea.*

This species of Walnut is not as valuable as the above; yet for its beauty, and the durability of its wood, it should claim a small portion of attention. The wood is rather soft for most purposes to which it otherwise might be applied. When grown near streams, or on moist side-hills, it produces regularly an ample crop of excellent nuts. It grows rapidly.

Shell-Bark Hickory—*Carya Alba.*

This, the largest and finest of American Hickories, grows abundantly throughout the State. Hickory wood possesses probably the greatest strength and tenacity of any other of our indigenous trees, and is used for a variety of purposes; but, unfortunately, it is liable to be eaten by worms, and lacks durability. For fuel, the Shell-bark Hickory stands unrivalled. The tree is ornamental, and produces every alternate year an ample crop of the best of nuts. It does not bear transplanting well, and the young trees should be preserved in their original situation wherever practicable.

Pignut Hickory—*C. Glabra.*

This species possesses all the good and bad qualities of the Shell-bark. The nuts, however, are smaller, and not quite so pleasant. It should be preserved and cultivated in common with the Shell-bark.

Red Beech—*Fagus Ferruginea.*

A common tree, with brilliant, shining, light-green leaves, and long flexible branches. It is highly ornamental, and should be cultivated for this purpose, as well as for its useful wood, which is tough, close-grained, and compact. It is much used far plane-stocks, tool handles, &c., and as an article of fuel, nearly equal to maple. The young trees do not bear transplanting well; those only of a few feet high can be removed with success. It is readily grown from seed, which had better be planted in the spring.

Water Beech—*Carpinus Americana.*

A small tree, called Hornbeam by many. The wood is exceedingly hard and compact, but the small size of the tree renders it almost useless.

Iron Wood—*Ostrya Virginica..*

This small tree is found disseminated throughout most of our wood-lands. It is, to a considerable degree, ornamental, but of remarkably slow growth. The wood possesses valuable properties, being heavy and strong, as the name would indicate; yet, from its small size, it is of but little use.

POPLAR.

The Poplars are trees of rapid growth. The wood is soft and white, and is not very useful. They thrive on almost any arable soil, and are readily propagated from cuttings.

Balsam Poplar—*Populus Candicans.*

This tree is of medium size, and is known by several names: Wild Balm of Gilead, Cotton Wood, &c. It grows in moist, sandy soil, on river bottoms. It has broad, heart-shaped leaves, which turn a fine yellow after the autumn frosts. It grows more rapidly than any other of our trees; can be transplanted with entire success when eight or nine inches in diameter; and makes a beautiful shade tree—the most ornamental of poplars. The wood is soft, spongy, and nearly useless.

Quaking Aspen—*P. Tremuloides.*

A well-known, small tree. It is rather ornamental, but scarcely worthy of cultivation.

Large Aspen—*P. Grandidentata.*

The largest of our poplars. It frequently grows to the height of sixty or seventy feet, with a diameter of two and one-half feet. The wood is soft, easily split, and used for frames for buildings. It is the most durable of all our poplars, and well worthy of our attention.

Sycamore, or Buttonwood—*Platanus Occidentalis.*

This, the largest and most majestic of our trees, is only found growing on the rich alluvial river bottoms. The tree is readily known, even at a considerable distance, by its whitish smooth branches. The foliage is large and beautiful, and the tree one of the most ornamental known.—The wood speedily decays, and when sawed into lumber warps badly; on these accounts it is but little used, although susceptible of a fine finish.—As an article of fuel it is of inferior merit.

Canoe Birch—*Betula Papyracea.*

A rather elegant and interesting tree. It grows abundantly in nearly every part of the State. The wood is of a fine glossy grain, susceptible of a good finish, but lacks durability and strength, and, therefore, is but little used in the mechanical arts. For fuel, it is justly prized, is to a considerable degree ornamental, and bears transplanting without difficulty. The Indians manufacture their celebrated bark canoes from the bark of this tree.

Kentucky Coffee Tree—*Gymnocladus Canadensis.*

This singularly beautiful tree is only found sparingly, and on rich alluvial lands. I met with it growing near the Peccatonica, in Green County. The wood is fine-grained, and of a rosy hue; is exceedingly durable, and well worthy of cultivation. It is readily grown from the seed, which should be immersed in boiling water for a few minutes, to insure a rapid germinaton.

June Berry—*Amelanchier Canadensis.*

A small tree which adds materially to the beauty of our woods in early spring, at which time it is in full bloom. The wood is of no particular value, and the tree interesting only when covered with its white blossoms.

EVERGREENS.

The Cone-bearing Evergreens—Coniferæ—are peculiarly interesting, as well for the invaluable qualities of their wood, as for their beauty, and the shelter and protection they afford, especially during winter, when all other trees are stripped of their green mantle. The seeds should not be taken out of the cones until they are to be planted, in April or May. A warm, sandy soil, which has a large proportion of decayed leaves, well incorporated into it, is the most suitable for raising seedling Evergreens.

WHITE PINE—*Pinus Strobus.*

The largest and most valuable of our indigenous pines. The wood is soft, free from resin, and works easily. It is extensively employed in the mechanical arts for a great variety of uses. It is found in great profusion in the northern parts of the State. This species is readily known by the leaves being in *fives.* It is highly ornamental, but in common with all pines, will hardly bear transplanting. Only small plants should be moved.

NORWAY, OR RED PINE—*P. Resinosa.*
YELLOW PINE—*P. Mitis.*

Two large trees, but little inferior in size to the White Pine. The wood contains more resin, and is consequently more durable. The leaves of both of these species are in *twos.* Vast quantities of lumber are yearly manufactured from these two, together with the White Pine, and yet the extensive pineries of the State are scarcely lessened.

SHRUB PINE—*P. Banksiana.*

A small, low tree; only worthy of notice here for the ornamental shade it produces. It is found in the northern sections of the State.

BALSAM FIR—*Abies Balsamea.*

This beautiful evergreen is multiplied to a great extent on the shores of Lake Superior, where it grows forty or fifty feet in height. The wood is of but little value. The Balsam of Fir, or Canadian Balsam, is obtained from this tree.

DOUBLE SPRUCE—*A Nigra.*

This grows in the same localities with the Balsam Fir, and assumes the same pyramidal form, but is considerably larger. The wood is light, and possesses considerable strength and elasticity, which renders it one of the best materials for yards and top-masts for shipping. It is extensively cultivated for ornament.

Hemlock—*A. Canadensis.*

The Hemlock is the largest of the genus. It is gracefully ornamental, but the wood is of little value. The bark is extensively employed in tanning.

Tamarack—*Larix Americana.*

This beautiful tree grows abundantly in swampy situations throughout the State. It is not quite an Evergreen, it drops its leaves in winter, but quickly recovers them in early spring. The wood is remarkably durable, and valuable for a variety of uses. The tree grows rapidly, and can be successfully cultivated in peaty situations, where other trees would not thrive.

White Cedar—*Cupressus, Thyoides.* (?)
Arborvitæ—*Thuja Occidentalis.*

These two trees are indiscriminately called White or Flat Cedar. They grow abundantly in many parts of the State, the latter in the northern section. The wood is well known as being exceedingly durable, furnishing better fence posts than any other tree, excepting the Red cedar. Shingles and staves of a superior quality are obtained from these trees. A beautiful evergreen hedge is made from the young plants, which bear transplanting better than most Evergreens. They will grow on most soils, if sufficiently damp.

Red Cedar—*Juniperus Virginiana.*

This is the well-known tree that furnishes those celebrated fence posts that " last forever." The wood is highly fragrant, of a rich red color, and fine grained; hence it is valuable for a variety of uses. It should be extensively cultivated.

There are many shrubs and vines indigeneous to the State that are worthy of notice for ornamental purposes, but it is not our purpose to speak of them here.

TREES NOT FOUND IN WISCONSIN.

There are many trees that have not yet been found growing in the State, the introduction and culture of which, we would especially recommend. Among these are :

Cucumber Tree—*Magnolia Acuminata.*

One of the largest and most beautiful of American trees. The wood is soft, light, and valuable for many purposes; but principally on account of its ornamental grandeur, we would recommend its introduction and cultivation.

Tulip Tree—*Liriodendron Tulipifera.*

This grand and noble forest tree is found in great profusion in Michigan, Indiana, and most of the adjoining States, where it is called White Wood or Poplar. The wood is of a yellowish color, easily worked, and sufficiently close-grained to admit of a good polish—strong enough for most uses, even where a considerable degree of strength is required. It is used extensively in the mechanical arts for a great variety of purposes, and grows rapidly. It is a beautiful tree, especially when covered with its large, tulip-like blossoms. It flourishes best in a rich alluvial soil.

Chestnut—*Castanea Vesca.*

A well-known, large and valuable tree, of remarkably rapid growth. It flourishes best in light, sandy, or gravelly soils. The wood is one of the most durable, capable of resisting a succession of heat and moisture for a considerable length of time; a valuable quality, which renders it especially suitable for fencing. In favorable situations, this tree produces an abundant crop of delicious nuts.

There are many other trees worthy of introduction, but to the three species mentioned, we would more especially call the attention of arboriculturists. Should either of these be known to grow within the State, we should be glad to be informed of the locality.

Locust—*Robinia Pseu-dacacia.*

One word in relation to the Locust, so much cultivated as an ornamental tree. Its rapid growth and durable wood, is all it has to recommend it. It is the last tree to put forth in the spring, and the first to shed its leaves in the autumn; which wither and fall without displaying any of those dying beauties so charming in many of our indigenous forest trees. The Locust is liable to the attack of the "borer;" an insect which is extensively destroying it in the Eastern States. The branches are so fragile, that the tree is frequently blown literally to pieces by the wind; to obviate which, to a considerable degree, we should "shorten in" all long and slender limbs. We cannot recommend its extensive cultivation, when so many better and more profitable trees can easily be obtained.

RAISING FOREST TREES FROM SEED.

In preparing seed-beds, a well drained, light, rich, sandy loam is best for nearly all trees. Dig deep, at least eighteen inches, and prepare as carefully as you would for a bed of choice vegetables. Plant in rows, three or four inches apart one way, and two feet the other. It is of great advantage to cover the surface of the ground, after planting, with decayed leaves, in imitation of nature. By this mulching, the ground will be kept moist, and thus facilitate the germination of the seed. The seedlings should be transplanted in the nursery, when one or two years old; six or eight inches by three feet is the proper distance in the nursery. It may, perhaps, be as well to plant the seed in the nursery at once, and save the first transplanting.

PLANTING.—In a timber plantation, the proper distance that trees should be planted from each other, must vary with the species and size of the young tree. The medium distance for trees of four or five years' growth, is four feet each way. It must be remembered, that many more must be planted than can grow to any great size. The object is to plant the trees so close that they will mutually afford protection to each other from the sun and storm. They must be thinned out from time to time, to give room for the most valuable, as they advance in size.

TIME FOR TRANSPLANTING.—All deciduous trees—those that shed their leaves in autumn—may be removed any time when the leaves are off. Early spring is generally to be preferred. Evergreens are more successfully transplanted, when the new shoots are just springing, which, in this climate, is about the first of June. Choose a damp, cloudy day, if possible; otherwise cover the roots from the sun with straw, moss, or matting. This precaution is indispensable when moving Evergreens, for if the fine rootlets ever become dry, the tree will surely perish.

PREPARATIONS FOR PLANTING.—Dig the holes before you procure the trees, in order to avoid unnecessary delays, for the sooner the trees are planted after they are dug up the better. Dig the holes for large trees six feet in diameter, and eighteen inches deep—smaller ones in proportion. Deposit the surface soil by itself, and if the sub-soil be poor, procure enough of good to replace it, so as not to be under the necessity of using this worthless soil dug from the bottom of the pit. Have ready good substantial stakes, sufficiently stout to effectually prevent the trees from motion. Do all this before you start for your trees, and not leave it to be done when your "poor trees, like so many fish out of water, are panting and suffering for a return to their native element."

Selecting and Digging up.—Procure trees of low and rather stalky growth, from open grounds, if possible. Avoid those of tall and slender form growing in deep and shady woods. Be sure the trees are healthy, young and growing. Having selected your tree, dig a trench two or three feet from the body all around it, deep enough to cut off every root; then dig under until it is easily loosened and turned out of its bed. Be careful and preserve every small root within the circle, for upon the preservation of the rootlets principally depends your success. Lay the trees upon a long wagon; protect the trunks and roots from being bruised and barked; cover the roots if the sun shines, or if there is a drying wind at the time.

Trimming and Setting out.—Cut off all the bruised and broken roots; next, cut off not less than one-half the top by shortening some, and cutting other limbs entirely away, as the shape of the tree may require. Set your trees no deeper than they originally grew, unless the soil be light and sandy, when you may set them three or four inches deeper. Having adjusted the roots in their natural position, drive your stakes before the roots are covered, that you may avoid injuring them; then fill up with finely pulverized surface soil, gently pressing it in every cavity. After the roots are lightly covered, pour in one or two pails-full of water; then finish by filling and treading it firmly.

Mulching.—Take half a bundle of straw to each tree; if the tree is of a large size, spread it evenly about the roots, then cover slightly with soil, leaving it a little "dishing;" this will prevent the evaporation of moisture, and keep the soil light and porous, permitting the water to penetrate freely. Old tan-bark or rotted chips will answer quite as well for this purpose as straw. You must not water your trees too often. Once a week, even during a protracted drought, is enough; daily watering is positively injurious. Do not consider the work completed until the trees are secured from motion, by firmly fastening them to the stakes; for be assured, if they are permitted to move about with every wind, they will perish. Transplanting trees in winter, with a ball of frozen earth attached to the roots, is the most successful method of removing large trees.

CONCLUDING REMARKS.

It is truly lamentable to see how much time and money is expended in planting trees, which for the want of information as to the *what, how,* and *when,* have already perished, or are in a most unfavorable state of decline, manifest by the sickly foilage, denuded trunks, or already dry and withered stems. And how could it be otherwise, since the trees have been kidnapped—forcibly taken from their damp and shady forest homes—the roots sadly mutilated, and then their long and slender trunks cut in two in the middle, and one end, (but little matter which) rudely thrust into a "post-hole" in the ground, and there subjected to the scorching rays of a midsummer sun; and then, as if to add insult to injury, they are asked to live, grow, and reward the perpetrator of all these outrages against vegetable life, by a luxuriant, healthy shade, and lovely ornament! It has been truly said, that "judicious planting, and the skilful culture of plantations, combine national and private interest in an eminent degree; for, besides the real or intrinsic value of the timber or ostensible crop, with other produce of woods, available for the arts and comforts of life, judicious forest-tree planting improves the general climate of the neighborhood, the staple of the soil as regards the gradual accumulation of vegetable matter; affords shelter to live stock; promotes the growth of pasture and of corn crops; beautifies the landscape, and thus greatly and permanently increases the value of the fee simple of the estate and adjoining lands." There is no country to which these remarks would apply with more truth than to this. How naked and cheerless those dwellings situated on the broad prairies, without a single tree to enliven the scene, and speak of shady comforts; but we hope to see a different picture when the farmers turn their attention more to the comforts and luxuries of Home. Finally, we say, plant trees young and old, and recollect, "What is worth doing at all, is worth doing well."

Yours very faithfully,

P. R. HOY.

To ALBERT C. INGHAM, Esq.,
    Sec. of the Wis. State Agr. Society.

LAYING OUT GROUNDS, FLORICULTURE, &c.

Milwaukee, December 10, 1852.

Dear Sir—It is with great pleasure that I comply with the request with which you honored me, to furnish a few ideas on the above subjects, for the use of the State Agricultural Society. Deeming these to be highly subsidiary to the great objects which the Society has in view, I only wish the task had fallen into more competent hands. However, I trust my attempt may at least have the effect of stimulating others, who have the capability to espouse the cause, and treat it more thoroughly and scientifically, on a future occasion.

Laying out Grounds.—Under this head may be comprehended all the operations requisite to the production of a finished residence, embracing the site for building, the embellishment of the grounds, and the judicious disposition of the fields, garden, orchard, &c. After the land is selected, the next important point is to choose the most eligible site for building. The mansion and offices demand peculiar attention, as they ought to form the centre of attraction, and present the principal feature of the homestead, for there are assembled all those scenes of usefulness, convenience, or elegance, which form the constituents of a country residence. In selecting the situation, a variety of circumstances, both of a local and general character, must be taken into consideration. Proximity, or otherwise, to the boundary line, or public road, the suitableness of the grounds contiguous for garden scenery, and trees, if any, and their capability of aiding in the general effect, belong to the former ; to the latter, the prospect from the house, the view of the house from a distance, shelter, facility for drainage, &c. A pleasant aspect for the principal rooms is always desirable, and in the absence of any local objection, perhaps that of the south, or south-east, would be the best. A south-west aspect is objectionable from the constant dazzling rays of the sun throughout the greater part of the day in summer, and a west aspect from the prevailing icy winds of winter. A north aspect for the front part of the house would at all seasons be too gloomy, although the view from the windows in that direction might be the most pleasant, as all vegetation looks the most luxuriant on the sunny side. A mere square, or oblong house, can, therefore, have only one really good aspect, and it naturally becomes a question as to what style of architecture would best secure a variety of aspects, consistent with the irregularity of offices, and other necessary appendages to a country mansion. The spirit of the Gothic style, more than any other, admits of this irregularity; and hence, the prevalence of that style in modern country residences. Among the advantages of this irregular style, one is, that it readily admits of additions in almost any direction, without compromising the character of the *tout ensemble*. When, however, a rigid adherence to the principles of good taste is aimed at, this style may not be always in keeping with the character of the neighboring grounds or surrounding country. Perhaps the more regular style of the Roman, or the simple style of the Grecian, would be more in place. In this case, the offices and erections for farming purposes might better form a distinct and distant feature in the landscape, or otherwise be concealed by trees, shrubbery, or walls behind the house.

The Lawn.—Whatever may be the style of the house, it is almost indispensable to a good residence to have a greater or less breadth of lawn extending in different directions from the principal front. This may be comprehended in one, or a series of terraces, or on a level, according to the position of the house and taste of the owner. Thanks to modern taste, and modern gardening, it is no longer considered necessary to have a square, or quadrangular hedged-in, or walled-in depository of plants to be called a garden. The whole lawn, or a portion of it, may be made available for a flower-garden, and will afford scope for a display of taste either of the most simple or the most elaborate—but this part of the subject I shall resume in its proper place.

At this juncture, it is to be presumed that the woodman has spared many trees, either single specimens, or groups, or both—for that reckless demolition of every tree and shrub round many settlements is lamentable to behold, and cannot be too highly reprobated. Well may they be called *clearings;* to-day, the noble denizens of the forest proudly wave their heads in stately grandeur, impressing the beholder with the power and majesty of nature; but to-morrow, alas, reveals another scene ! The ruthless age has robbed fair nature of her ornaments, and left instead a wilderness of ghostly stumps, guarded on every side—as if in very mockery of their stability—with bristling lines of zig-zag fences. Here, amidst this desolate array of stumps and rails, a family of young children have to be reared, and what else can be expected but that the minds of the latter will be as destitute of ideality and refinement as the aspect of the former is of comeliness. After a few years have elapsed, an embryo orchard shoots up to relieve the general monotony, distinguished by a number of green lines, so formally and abrubtly defined that it looks like a thing *per se*, having no connection or harmony in itself, or with anything else in nature.

Few States in the Union can compare with Wisconsin in the capability of furnishing materials for creating beautiful landscapes; nor need we go out of it to seek for the choicest models. The world-renowned parks of "Old England" do not afford better opportunities than our own oak openings—with their graceful and flowing outlines, and their gently undulating sweeps, the noble vistas—here contracted, there extended—in this direction, radiant with some gleaming lakelet half concealed; in another, reflecting sombre shadows from some lazy stream; and as a whole, producing such harmony of coloring, so rich a distribution of light and shade as cannot fail to delight, but never cloy, the most artistic eye.— When a dwelling is located amidst the oak openings, the occupant can have no difficulty in laying out his grounds, for the scenery around will be suggestive of the character of all his out-door embellishments, and his aim will be to make them tend to the idea of connectedness and consistency with that. In such situations, to make startling contrasts is attended with no small amount of expense, and a demand of no commonplace skill. In this State, nature is so lavish of the materials for verdant decoration that little or no expense need be incurred in furnishing, but rather in thinning out, and only subduing a reasonable portion of the land for usefulness or convenience, without impairing any of the natural or local beauties. When it is desired to give a park-like appearance to a whole farm, where the land is well timbered, nothing can be more easy. The square, or quadrangular form, may generally be the most convenient for fencing off the various fields, but it is not necessary that they should assume that form. By a little management, the fences at a certain point may be so concealed by clumps of trees left standing as to take away any formality, and obscure the real boundary of a field. The size of the fields, and the distance and distribution of the clumps, will of course depend on the extent of the farm, and the purposes to which it is devoted. The clumps of trees should consist of irregular, detached masses, or occasionally single trees, when they possess individual beauty to recommend them, and can maintain the idea of connectedness. Under this arrangement, a farm would partake of the character of a forest scene subjected to the plow; and if extensive, and really forest-like, resemble those woody districts in Germany where "cultivation smiles in the glades and recesses of eternal forests."

FENCES.—Of all kinds of fences, the common rail or zig-zag fence presents to the eye of strangers the most uncouth appearance, as well as that of a prodigal waste of the "raw material." Perhaps they may be, in many instances, the most economical, as regards cheapness of construction, strength and durability; but where facilities exist for sawing timber, these qualities might be made to combine with a more agreeable form, and a tithe of the amount of material. The improved state of the manufacture of iron, affords new and desirable accommodations in the way of light, cheap and durable fences, desiderata of no small importance in districts where wood is not abundant.

PLANTING.—When a necessity exists for planting young trees, either near or remote from the house, as a general rule, a selection from those found in the neigborhood would be best; as they will not only better harmonize with those around, but grow and thrive with more certainty than others, transferred from a different soil and climate. As to the best kinds for planting on the rich, black soil of our prairies, I have had no opportunity of forming an opinion; but I should expect that most of the soft-wooded, rapid-growing kinds would succeed. As fencing must be a comparatively expensive undertaking on extensive prairie lands, it would be well for settlers thereon to provide themselves at the outset, with seeds of such plants or shrubs as will most rapidly or permanently afford them live fences or hedges. Of these may be mentioned the Osage Orange, Locust, Buckthorn, Hawthorn, Beech, &c. It is affirmed by those who have had experience, that the Osage Orange will form a hedge fence in five years from the time of sowing, capable of resisting all sorts of cattle, and at the same time so compact as to produce a complete barrier, even to rabbits and other small vermin. That this rapid growing tree will survive the winters of this latitude, has been proved beyond a doubt within the State, and in several instances. One fine, thrifty tree may be seen in this city, in the garden of Lewis Potter, Esq., which has stood the test of over ten winters.

THE VEGETABLE GARDEN.—This necessary appendage to a family mansion, should, when circumstances permit, be formed convenient to the rear of the building, out of the view from the principal rooms, and the front approach. A very convenient locality would be such as to have easy access to the domestic offices for culinary purposes, and to the stables and farm buildings for manure.

THE ORCHARD.—The situation of the orchard—all other circumstances being suitable—might be near the garden; and where a gardener or overseer is kept, his house, together with the fruit house, and other receptacles for the winter storage of roots and vegetables, might be very properly placed between them.

THE ROAD, OR APPROACH.—When of any material length, the approach should be formed so as to reveal gradually any natural or artificial beauties the place affords, and each turn should be produced by some gentle variation of the surface, a clump of shrubs or single trees. The first or most distant view of the house, should be as favorable as possible; and the nearest or close view, to comprehend the front entrance or porch, together with the most refined creations of ornament in its vicinity, at a glance.

WALKS.—These are often necessary accompaniments to home scenes that cannot otherwise be seen except at certain seasons and conditions of the surface. Walks should always bear some degree of analogy to the scenes they pass through. Their course should be dictated by the range of attractions to be seen, and their turnings by some local or accidental beauty, being careful that there should be a sufficient reason for such arrangement.

THE FLOWER GARDEN.—There are perhaps few departments of out-door pastime surrounded by more delightful and refining associations, than the cultivation of a tasteful and elegant flower garden. It is a field equally open and inviting to the humblest cottager and his wealthier neighbor, as well a source of agreeable recreation to the son of toil as to the heir of luxury. The flower garden is a never failing concomitant of a refined civilization; and the entire absence of flowers around a dwelling, bespeaks either abject penury or a rude ignorance of the sweets of life. In the gloomy retreats of forest life, what can be more refreshing and delightful than the companionship of flowers? Though all else around be new, in these we can recognize the sweet companions of our youth; the silent, yet eloquent souvenirs of our early home and father-land. If parents desire to cultivate in children their powers of fore-thought and patience, let them give them a flower garden. The impressions received there will be as lasting as life itself; and the maxims employed to make them successful florists, will not be lost on their character and moral bearing in after years. It is rarely that a parent is found fond of his plants and negligent of his child; or gentle towards his flowers, and rough towards his own kind. Hence, we might expect that the cultivation of flowers will favor the amenities of domestic life, and that a wide spread taste and interest in Floriculture, will be a remedy for idleness and other demoralizing habits.

The flower garden, as I suggested before, might be consistently formed on the lawn or terrace, in front of the house. The usual way is to cut out certain figures in the turf, each figure or set of figures, to bear, in size, a just proportion to the extent of the lawn, and the magnitude of the house. In order to be effective, they should be symmetrical and have one general character of outline, whether of straight, curved, or composite lines; nor should their size differ so much as to give the idea of large and small ones mixed together. The figures, or system of figures, however detached, should always be formed to harmonize with some obvious design, and show decidedly their connection with a general centre. Sharp angles or projecting points, should be avoided, as such parts cannot be properly covered with plants, and must always have a bad effect. Neither should the surface of the beds be much raised above the level of the turf, for in such case they look blotchy, and the plants in them too much subjected to the influence of drought for their health and vigor. In planting the beds each may contain flowers only, or flowers and shrubs mixed; but a better effect is produced by each having only one kind of flowers, or one kind of shrubs, so as to form masses, or shades of color to harmonize with those of the other beds. Of all plants or shrubs, capable in themselves of forming an interesting and varied flower garden, the rose, perhaps, stands alone. It is to be found of nearly every shade of color, and by a judicious selection of kinds, a continued succession of bloom may be had from June till December. A very effective method would be, to have each bed filled with roses of one color, containing a mixture of summer and autumn blooming kinds, so that no bed would be without a perpetual display of flowers. A very neat method of keeping roses dwarfish and compact, is, to peg down the branches to the ground: this will, at the same time, expose a greater surface of the plants to the light, and induce an increase of foliage and bloom.